WATER MANAGEMENT AND SUPPLY

Paul N. Cheremisinoff

P T R Prentice Hall, Englewood Cliffs, New Jersey 07632

Library of Congress Cataloging-in-Publication Data

Cheremisnoff, Paul M.
    Water management and supply / Paul M. Cheremisinoff
      p. cm.
    Includes index.
    ISBN 0-13-501214-7
    1. Water quality. 2. Water quality management. 3. Water-supply.
I. Title.
TD370.C5 1993
333.91--dc20
                                                                                       9312555
                                                                                       CIP

Editorial Production Supervision: bookworks
Acquisitions Editor: Michael Hays
Cover Design: Jerry Votta
Buyer: Mary Elizabeth McCartney

 © 1993 by P T R Prentice-Hall, Inc.
A Simon & Schuster Company
Englewood Cliffs, New Jersey 07632

The publisher offers discounts on this book when ordered
in bulk quantities. For more information contact:

    Corporate Sales Department
    P T R Prentice Hall
    113 Sylvan Avenue
    Englewood Cliffs, New Jersey 07632
    Phone: 201-592-2863   Fax: 201-592-2249

All rights reserved. No part of this book may be
reproduced, in any form or by any means,
without permission in writing from the publisher.

Printed in the United States of America

10 9 8 7 6 5 4 3 2 1

ISBN 0-13-501214-7

Prentice-Hall International (UK) Limited, *London*
Prentice-Hall of Australia Pty. Limited, *Sydney*
Prentice-Hall Canada Inc., *Toronto*
Prentice-Hall Hispanoamericana, S.A., *Mexico*
Prentice-Hall of India Private Limited, *New Delhi*
Prentice-Hall of Japan, Inc., *Tokyo*
Simon & Schuster Asia Pte. Ltd., *Singapore*
Editora Prentice-Hall do Brasil, Ltda., *Rio de Janeiro*

CONTENTS

Preface   v

1 WATER SOURCES AND USES   1

    Introduction   1
    Sources of Water Supply   3
    Water Distribution   23
    Sources   27
    Water Pollutant Categories   37

2 WATER QUALITY AND PROPERTIES   40

    Goals of Water Treatment   40
    Water Composition   46
    Physical Properties   59

3 CHEMICAL PROPERTIES   64

    Hydrate Formation   66
    Solutions   69
    Colloidal Solutions   77
    Emulsions   79
    Ions and Ionization   80

4 EXAMINATION OF WATER   83

    Measurement Units   83
    Color   87
    Odor   88
    Taste   90
    Alkalinity   95
    Fluoride   97
    Hardness   99
    Residual Chlorine   105
    Hydrogen-Ion Concentration—pH   107

Bacteria of Coliform Group 109
Microscopic Examination of Water 116

5 WATER SUPPLY PURIFICATION   120

Aeration 122
Suspended Impurities 123
Flocculation 130
Sludge Blanket Clarifier 136
Filtration 137
Filter Media 143
Classifications 148
Disinfection 156
Fluoridation 163
Ion Exchange—Demineralization 165

6 GROUNDWATER   167

Withdrawal and Consumption 167
Groundwater Hydrology 167
Contamination 174
The Safe Drinking Water Act (SDWA) 178
Sole Source Aquifers 179
Groundwater Monitoring 180

7 INDUSTRIAL AND COMMERCIAL REQUIREMENTS   204

Flow Rates, Pressures, and Storage 205
Quality 205
Water Conditioning 206
Boiler Feed Water 209
Commercial and Institutional Water Conditioning 220
Corrosion 222

8 COOLING WATERS   236

Problems Inherent to Contaminants 237
Treatment of Cooling-Water Systems 244
Methods of Evaluating Cooling-Water Inhibitors 251
Langelier and Ryznar Equations: Saturation and Stability Index 251
Organic Growths 252
Legionnaires' Disease 253
Water Analysis and Treatment 253
Plastic Cooling Towers 259
Guidelines for Integrated System Operation 264

Index   265

PREFACE

Water is provided by nature so generously that most of us take it for granted and use it without ever considering how little we know about it. The three factors of life in order of importance are air, water, and food. Water was considered one of the four elements by the ancients with the others being earth, fire, and air. Without a plentiful supply of usable water, the whole structure of our society would collapse.

Water is the best of all liquid solvents. More substances dissolve in water to an appreciable extent than any other liquid. Because of this solvent property, naturally occurring water is never pure. It becomes contaminated with impurities in nature as well as through various human activities. Because of this, water in our world today requires adequate management and maintenance through understanding of the environment and judicious use of technology.

This book is written as an objective overview of water—its properties, sources, and uses in industry, commerce and transportation. The material presented hopefully will serve to summarize information on water as well as describe many of its unique properties. While we accept water as nature provides it in most cases, aquatic environments are as numerous as water itself and its supply and purity are continually compromised by both human and nature.

Common to all water from various sources are the differences from water habitats and origins. Additionally humans have been able to significantly impact the properties of water through their activities. The effects of pollution assume many characteristics and affect our water supply. As people use water or even as it is found in nature, properties such as color, turbidity, taste, odor, chemical pollutants and microorganisms must be taken care of.

Water supply management is of interest to the engineer, chemist, and manager because the business of the supply and treatment of water involves technical problems and water is a fundamental requirement in manufacturing as well as for municipal and public uses. The importance of a dependable water supply has been recognized since ancient times.

PAUL CHEREMISINOFF

1 WATER SOURCES AND USES

INTRODUCTION

The three vital factors of life in order of importance are air, water, and food. The availability of clean water and its use are at the foundation of our civilization. Without a plentiful supply of water the whole structure of our society would collapse. Today water requires adequate management and maintenance through understanding of the environmental and judicious use of technology.

WATER

Molecular weight	18.016	Boiling point	100°C
Critical temperature	+ 374°C	Freezing point	0°C
Critical pressure	217.5 atm.	Specific gravity	1
Formula (gas)	H_2O	Specific heat	1 cal

Water was considered one of the four elements by the ancients with the others being earth, fire, and air was not recognized that water is a compound substance until comparatively recently. In 1781, Lord Henry Cavendish was the first to prepare water by the combustion of hydrogen in air. It remained for the famous French chemist Lavoisier to show that water is composed of hydrogen and oxygen only.

Historically water has been consumed as an inexhaustible natural resource with little or no concern for costs, pollution, purification or transport. As late as 100 years ago, water for residential use came mainly from wells or other fresh water sources located near the point of use. Water use for agriculture was minimal with an abundance of prime land requiring only natural rainfall. Irrigation was primitive with diversion of small water streams. Industrial uses relied on nearby lakes or streams.

The availability of water is generally taken for granted in water-rich areas, and it is often assumed it can be found in an endless supply. It is available in large quantities from a variety of sources, although not distributed uniformly over time and geography. In recent times there has become an awareness of the delicate balance between water uses, availability, and costs.

Water is the most abundant and most widely distributed compound. As a solid in the form of ice and snow, it covers the colder regions of the earth. In the liquid state as lakes, rivers, and oceans, it covers about three fourths of the earth's surface, sometimes reaching a depth of nearly six miles. It is present in the air as a vapor with often as much as 50,000 tons of it in the air over a square mile of the earth's surface. It is present in all living matter; indeed almost 65 percent of the human body is composed of water. All our food materials contain water, the content varying from 7.3 percent in oatmeal to 94.7 percent in lettuce.

Water supply is of interest to the engineer, chemist, and manager because the business of supplying and treating of water involves engineering problems and because water is a fundamental requirement in many manufacturing and municipal uses. Among the many problems in this field of a primary concern is the nature of the sources, examination, and purification of waters.

The importance of a dependable water supply has been recognized since ancient times. The digging of wells dates back to early Chinese and Egyptian history and aqueducts of the ancient Romans are considered today to be remarkable engineering achievements. The existence of a dependable water supply lost importance during the Middle Ages, and it was not again recognized until the eighteenth century, when London and Paris began to adopt the principle of a water supply for individual houses. In these early domestic supplies, however, the water without purification was turned on for only a short period each day and such intermittent service was not entirely displaced until almost the end of the nineteenth century. There has been an enormous expansion since the advent of cast-iron pipe, mechanical pumps, and quality control. Today water of established purity is almost universally expected, supplied, and utilized.

The largest water requirement is for municipal use, but the purity standards for this purpose are sometimes quite different from those demanded for industrial and commercial use. For manufacturing purposes the quality of water may be so important as to be the controlling factor in the location of an industry. Knowledge of the character of available water supplies and the effects of water impurities on industrial processes is therefore essential to the choice of a plant's location. The following list provides some of the most critical location and selection factors for water supply:

- Water needs: process, cooling, potable, fire protection
- Water availability: public water supply, private water supply, ground water, surface water
- Groundwater geological potential (on site)
- Water characteristics: chemical, bacteriological, corrosiveness
- Water distribution: amount available, pressure, variations, proximity to site, size of lines

- Cost of water supply: extension of existing service, development of new supply, cost per 1,000 gallons
- Water treatment requirements: process, cooling, boiler feed water, potable, and other
- Special considerations: restriction on use, future supplies, compatibility for use in process

SOURCES OF WATER SUPPLY

Although rainfall is the ultimate source of all fresh waters, water supply is usually directly acquired from surface waters such as lakes and rivers or from groundwaters such as springs and wells. There are, of course, seasonal variations in the rainfall which are reflected in the quantity and quality of the water supply. The variations of rainy and dry seasons in the swollen turbid floods of springtime and the low water of late autumn are well known.

Natural Water Impurities

Natural fresh waters contain impurities and the nature and quantity of these impurities depend on the conditions under which the waters have accumulated. The value of a water supply is, therefore, closely associated with the nature of its source.

The impurities usually found in natural waters may be classified as suspended and/or dissolved. The former includes microscopic and bacteriological impurities. Suspended impurities produce turbidity and may range from none in a clear spring water to dense mud. Dissolved impurities usually consist of the carbonates, bicarbonates, chlorides, sulfates of calcium, magnesium, and alkalies, and may range from a few parts per million in snow water up to several thousand parts per million in mineral springs. Other soluble impurities such as silica, iron, alumina, nitrate, and so on, may be present to the extent of a few parts per million. Microscopic impurities include the varieties of plant and animal life above bacterial and may range in quantity up to such concentrations that they impart strong turbidity, color, taste, or odor to the water. Bacteriological contamination may include yeasts and molds as well as bacteria and is usually estimated both in terms of the total number of all species and the number of gas-forming organisms. The total number may range from a few per cubic centimeter in a deepwell water to thousands per cubic centimeter in a surface water. Gas-forming bacteria may range from none in a mountain stream to hundreds or even thousands per cubic centimeter in the polluted waters of a densely populated region. An important consideration for use is not the amount on hand but rather the rate at which it is being continuously renewed in the hydrologic cycle. Without this replenishment the available supply can soon be exhausted. The total amount stored is important since it enables us to draw water continuously even though the replenishment does

4   Water Management and Supply

not take place at a constant rate. Over the long term we cannot use water at a rate which exceeds the rate of replenishment.

In areas where a storage capacity is available, the runoff which occurs during wet periods can be stored and saved for use during dry periods. Most of the naturally stored surface water is contained in a small number of very large lakes which are geographically scattered. In most areas it has been necessary to add to the natural storage capacity by the construction of impounding reservoirs. It has also been necessary to transport water to areas where the supply is not adequate. Figure 1-1 shows an impounding reservoir.

Figure 1-1 Impounding reservoir

The Hydrologic Cycle

The science of hydrology is concerned with the distribution of the waters of the earth and with the endless movement of these waters from the sea to the atmosphere, from the atmosphere to the land, and from the land back to the sea. This process known as the hydrologic cycle, is shown by Figure 1-2. The oceans cover 71 percent of the earth's surface and contain over 97 percent of the earth's water. The other 3 percent is found in the atmosphere as water vapor, on land as fresh water, snow, and ice, and beneath the surface of the land as groundwater.

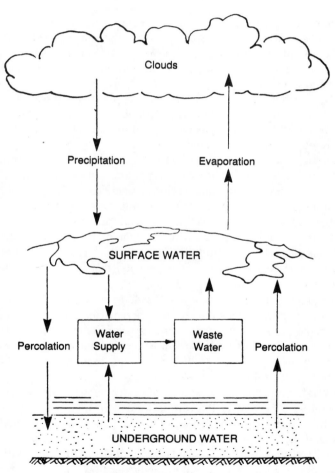

Figure 1-2 Hydrologic cycle

Not all water moving through the hydrologic cycle spends the same time completing a cycle. Water stored in the polar ice caps and glaciers and water stored in deep aquifers (layers of rock, sand, or gravel) may have been there for thousands of years. On the other hand, water flowing in a river at any given time may have been in the ocean only a few days earlier. About three fourths of the fresh water is more or less permanently stored in the polar ice caps and glaciers; most of the rest of it stays in soil and rock below the surface of the land. Only about one third of 1 percent of the total fresh water is in lakes and rivers—and about 20 percent of this is in the Great Lakes between the United States and Canada. The average time required for fresh surface water to complete the hydrologic cycle may be on the order of three to four years, due largely to the long holdup in the large lakes. Most of the precipitation which falls on land never reaches the ocean but completes its cycle back to the atmosphere by the processes of evaporation and transpiration (the movement of water from the soil to the atmosphere through growing plants). For this portion of the fresh water, the cycle is completed in hours or days.

About one fourth of the earth's fresh-water supply is stored beneath the surface of the land where it remains for hundreds or thousands of years. Only a relatively small proportion of this water is located in soil formations (aquifers) from which it can be drawn in significant amounts.

The hydrologic cycle is the process whereby water is converted from its liquid or solid state into its vapor state. As a vapor the water is capable of traveling considerable distances from its source prior to recondensing and returning to earth as precipitation. The hydrologic cycle is a complex, interrelated system involving the movement of atmospheric, surface (marine and fresh), and groundwater throughout various regions of the world. It is the hydrologic cycle that is solely responsible for the world's precipitation, and it is this precipitation, falling on the terrestrial and surface fresh-water environments, that is the sole source of the earth's supply of fresh water (Figure 1-2).

The hydrologic cycle may consist of either long or various short cycles. In the short cycles, water may evaporate from either marine or fresh-water systems, condense almost immediately, and return as precipitation to the same system. Another variation of a short cycle is the precipitation and subsequent evaporation of water from land surfaces, followed by its condensation and return as precipitation to land, followed by reevaporation, and so on, repeating the cycle.

In the overall cycle the major source of water vapor is the world's oceans, which contain 97.3 percent of the earth's waters. In this cycle a portion of the water evaporates to form clouds that move inland. The water vapor then cools and returns to earth as precipitation. It is estimated that only 0.007 percent of the oceanic water is distributed to terrestrial areas annually. This water will ultimately return to the oceans through river and groundwater flow. Since precipitation may occur close to the source of initial evaporation or thousands of miles away, the water may remain in the vapor state for variable times such as a few hours to a few weeks. The average residence time for water to remain in the atmosphere is considered to be ten days.

Evaporation and Condensation

Evaporated primarily from the oceans, water vapor tends to condense around minute particles, termed *nuclei*, which are suspended in the atmosphere. The nuclei generally consist of small particles of organic material such as spores, pollen, fine mineral particles, volcanic ash, and the like. Dust and smoke particles from industrial sources and automotive exhausts may also serve as nuclei and are a major factor in contributing to the contamination of rainwater. This is due to the fact that airborne particles (such as lead) readily dissolve in the newly condensed atmospheric water, and these materials are then returned to earth along with the rainwater.

Initially, the condensed liquid is in the form of extremely small droplets, that is, less than 0.04 millimeter (mm) in diameter. Because of their small size, their rate of fall is negligible, and they are retained in the atmosphere as clouds. Ultimately, however, this vapor forms precipitation. The largest amount of precipitation forms when masses of warm, moist air move into regions of cold air resulting in a rapid condensation of the vapor and subsequent precipitation. Another factor arises during periods of warm weather when air, warmed at the earth's surface, decreases in density and rises into overlying cold air, bringing about condensation of the water vapor. A third mechanism involved in condensation occurs with the cooling of air masses as they move over high mountains. In each case the cooling of warm, moist air is responsible for the condensation and precipitation of rainwater.

Runoff/Stream Flow Infiltration

When rain falls on terrestrial areas, a portion of the water is caught by vegetation. This process is termed *interception* and this water is readily reevaporated, since there is a large surface area that is exposed to wind action. The remainder of the water falls to earth, and a portion sinks into the soil surface by a process known as *infiltration*. The portion not infiltrated into the soil, termed *surface runoff*, flows over the surface and is discharged in undeveloped areas into streams. The water entering a stream by surface runoff plus the water entering via groundwater flow is termed *runoff*. Thus the terms *runoff* and *surface runoff* are different and distinct. Surface runoff equals precipitation minus the water lost by interception and infiltration. Runoff, on the other hand, is generally synonymous with stream flow and is the sum of the surface water plus the groundwater that enters a stream.

Infiltration is also distinct from groundwater because when the water percolates into the earth's surface, different portions will follow three distinct pathways. A portion of the water will function as interflow because the presence of impermeable lower sediments will prevent deep penetration of this water. Consequently, this segment will flow just below the soil surface and discharge into streams. Another portion will remain above the water table in an area termed the *zone of unsaturated flow.* Both of these portions of water are considered to be in the zone of aeration. The third portion will percolate down into the groundwater table and will eventually be discharged into streams.

The passing of water through soil soluble material (for example, gypsum beds) will increase the dissolved-solids content. In order to express this increase in a quantitative manner, an equilibrium-type relation is usually assumed and the amount dissolved depends on the degree of unsaturation of that particular species. For example, in normal irrigation practice leaching is used for accumulated-salts removal and to maintain an adequate salt balance. The leaching process produces a continual change in reused water quality. Figure 1-3 shows irrigation practice for crop management.

Figure 1-3 Irrigation practice for crop management

Dispersion of fluid streamlines or water particles during flow through porous media is characterized by differences in fluid velocity as the water flows through cracks and around granules. The concentration front of a tracer material or pollutant traveling down a column or through an aquifer usually develops a one-sided normal distribution due to this dispersion process. The degree of dispersion in porous media is characterized by a dispersion coefficient. Attempts to predict this coefficient for field situations have been totally unsuccessful. For well-sorted sands under carefully controlled laboratory conditions excellent results have been obtained, but when applying such results to large-scale field experiments errors of two to three orders of magnitude have resulted.

Evapotranspiration

A large amount of the precipitated water is converted to vapor by evaporation and/or transpiration. Evaporation is where molecules of liquid water at the surface of a water body or in moist soil gain sufficient energy to leave the liquid state and enter the vapor state. The energy absorbed by the water in this process is stored within the vapor--hence the large amount of heat contained in the moist air.

Transpiration

Transpiration is the process whereby terrestrial and emergent aquatic vegetation release water vapor to the atmosphere. In photosynthesis plants take in liquid water and carbon dioxide and, by a complex series of reactions, convert these materials to carbohydrates, oxygen gas, and water vapor. Submergent aquatic vegetation where plants grow completely beneath the water surface release water and oxygen as vapors but, in these cases, the oxygen and water vapor produced and released into the surrounding liquid water will immediately form hydrogen bonds with the liquid water. Water vapor will be converted to its liquid form, while the oxygen remains in the water column as dissolved oxygen. Since the leaves of both terrestrial and aquatic emergent vegetation are in air, the water vapor and oxygen, when released by these plants in photosynthesis, will remain in the vapor state because there are insufficient water molecules in the immediate vicinity to form hydrogen bonds and place the water vapor and oxygen in solution.

Water converted from liquid to vapor in transpiration is considerable, and the ability of plants to remove water from both soil and aquatic systems cannot be underestimated. Transpiration by emergent vegetation is a major factor in the "drying up" of lakes.

Groundwater, unless it is within a few feet of the surface, is not evaporated. Rather, the portion that is converted to vapor is transpired by plants. In most regions the water lost by evaporation cannot be measured separately from the water lost by transpiration. Consequently, the two are considered together as evapotranspiration. The most lengthy portion of the hydrologic cycle is completed when groundwater is returned to the earth's surface. The return may occur by springs, transpiration, or by artificial means. Any natural surface discharge of sufficient water that will flow as even a small rivulet is termed a *spring*, while a smaller discharge is called *surface seepage*. Groundwater may also be discharged as subaqueous springs below the surface of lakes, rivers, and marine systems.

The earth's waters are, therefore, interconnected by the hydrologic cycle. Oceanic water is evaporated and a portion is carried as water vapor over terrestrial regions, where it returns to earth as precipitation. This water may then be evaporated or transpired, or it may enter streams, rivers, lakes, or the groundwater system. Regardless of the pathways that the water may take, it is the only source of fresh water and, therefore, the only source of domestic and industrial water. Ultimately, and generally only after use and reuse by people, this water will return to the sea—only to be evaporated to reenter the hydrologic cycle.

Surface Water Supplies

Although about three fourths of the public water supply systems in the United States are from underground sources, these systems supply only about one fourth of the people who are served by public systems. Most large cities are dependent on surface supplies.

The variety of substances determining surface water quality include inorganic, organic, and suspended matter, and products of the biologic system. These substances respond in different ways to natural and man-made environments. The water quality of a stream is determined by its environment—climate, geologic, hydrologic, physiographic, biological, and cultural. Waters of similar quality can be expected from terrains of similar environments, and useful correlations between water quality and environment are probable. Some areas display a consistent pattern of overall similarity, but with an internal pattern of variability that gives a wider range of concentration than one would expect. This makes the setting of firm standards of natural quality for river segments and river systems difficult.

Table 1-1 includes a cross section of parameters to be considered for any comprehensive data collection program. Parameters to be observed in any basin should be selected with reference to federal, state, and local agencies to form a manageable number of base or index of measurements. Although the parameters listed in Table 1-1 do not constitute an all-inclusive program, they include the important factors involved in water quality management. Furthermore, they comprise the parameters to be included in river quality standards that have been adopted by regulatory bodies for pollution abatement. These base line parameters are divided between so-called conservative (stable) and nonconservative (unstable) parameters and emphasize basic data of utmost importance for the overall assessment of inorganic, organic, and biological stream quality.

An instantaneous water quality measurement provides current information on concentration at a given place and time. The instantaneous measurement may be a field determination, a laboratory determination, or a value sensed and recorded

Table 1-1 Water Quality Parameters

Major Parameters	Parameters
Eations and anions	Salinity Inorganic variability
Specific conductance of dissolved solids Minor elements Hardness (Ca + Mg)	Geochemical
Nitrogen cycle (NO_2, NO_3, total N)	Nutrients
Phosphorus cycle (PO_4 and total P)	Nutrients
Dissolved oxygen	Oxidation Aquatic health
pH	Acidity and close-up corrosiveness Geochemical Aquatic health
Radioactivity gross alpha, beta	Radiochemical
Carbon total or dissolved C	Organic
Pesticides	Organic Ecology, public health
Phytoplankton and benthic organisms (selected)	Biological integration
Turbidity or suspended sediment	Physical
Temperature	Physical
Color, odor	Physical
Biochemical oxygen demand (BOD)	Organic waste
Microorganisms (coliform)	Waste, human and animal

by a monitor. It must be within verifiable limits, especially for meeting legal requirements associated with stream quality standards. It also provides information for many other management requirements and uses.

Rainwater

Rain, the purest of natural waters, may collect finely powdered mineral matter, particles of plants, ammonia, carbon dioxide, oxygen, nitrogen, and other substances gathered in its fall through the atmosphere. On reaching the earth, rain flows into streams as surface runoff, passes into the air as vapor, or sinks into the ground, a storage reservoir from which it eventually emerges as vapor or surface water, or is recovered as groundwater.

Precipitation and Runoff

In the United States the average annual precipitation is about 30 inches, of which about 21 inches are lost to the atmosphere by evaporation and transpiration. The remaining 9 inches become runoff into rivers and lakes. Both the precipitation and runoff vary greatly with geography and season. Annual precipitation varies from more than 100 inches in parts of the Northwest to only 2 or 3 inches in parts of the Southwest. In the northeastern part of the country, the annual average precipitation is about 45 inches, of which about 22 inches become runoff. There is some variation from place to place and considerable variation from time to time. During extremely dry years, the precipitation may be as low as 30 inches and the runoff less than 10 inches. In general, there are greater variations in runoff rates from smaller watersheds. A critical water supply situation occurs when there are three or four abnormally dry years in succession.

Surface Waters

The term *surface water* is applied to the natural water in streams, ponds, and lakes, although it is understood that such waters may have percolated more or less through the soil in their course to the open streams. Actual surface drainage is the chief contributor of turbidity to streams, while the percolated water is responsible for a major part of the dissolved mineral impurities. Whether passing over the surface or slowly percolating underground, water takes into solution a part of the materials with which it comes in contact. Impurities derived from the air assist in the solution of rock-forming minerals, while all the dissolved materials exert an influence upon each another. The chemical constitution and physical conditions of an area are therefore reflected by the water it furnishes.

The amount and intensity of rainfall, assisted by the surface slope and the permeability of soil, play an important role in the quality of water. Other things being equal, the least mineral matter is found in waters from regions of the highest rainfall. In periods of drought, groundwater is the chief source of stream flow and consequently, the stream waters are thus concentrated. A gentle rain will increase the groundwater and cause it to feed the streams more generously but will not contribute appreciably to stream flow by means of surface runoff. A hard rain, on the contrary, may contribute comparatively little to groundwater but will wash much dirt into streams, increase their loads of suspended matter, and dilute their dissolved mineral impurities. The amount and character of impurities in surface water, therefore, depend on numerous conditions and are derived from many different sources.

Safe Yield of Surface Water Supplies. One inch of runoff per year from a watershed area of 1 square mile is equivalent to a stream flow of 0.0737 cubic feet per second, or 0.0477 million gallons per day (mgd). In extremely dry years the average runoff from many watershed areas may be less than half of this. The actual momentary rates of flow vary over a very wide range.

In order to be able to draw water continuously at or near the average rate of flow from a watershed, the water must be stored during periods of high flow. The safe yield of a watershed area is the maximum rate at which the water can be withdrawn continuously over a long period of time. Without storage, the safe yield would be equal to the minimum rate of flow to be expected in the future.

Quality of Surface Water Supplies. In general, the quality of surface water from streams, lakes, or impounding reservoirs is unsafe for human consumption and requires treatment. Small, spring-fed upland streams may yield practically clear and tasteless water, except during the height of rainstorms when a moderate amount of suspended solids may be present. Though any objectionable bacteria present may be of animal origin, such streams are always exposed to accidental or incidental pollution by humans.

Large streams usually drain inhabited watersheds and receive serious pollution by surface drainage from eroded or plowed fields so that the physical character of the water is usually inferior to that from brooks which drain wooded areas. Inadequately treated sewage and industrial wastes may be discharged directly into many streams. Lakes, ponds, and impounding reservoirs yield water of better quality than most flowing streams because of the beneficial effect of self-purification by sedimentation and storage.

Rivers. Water supplies from rivers normally require more extensive treatment than those from other sources. The turbidity, mineral content, and degree of pollution will vary considerably from day to day, while the temperature may vary considerably throughout the year.

Usually a river supply is chosen only when other dependable sources are not available. A supply drawn directly from a river has an advantage over a supply drawn from an impounding reservoir in that original costs are less because expensive dams, conduits, and purchases of extensive land and water rights are not necessary.

Natural Lakes. Large lakes may yield water of exceptionally fine quality except near the shoreline and in the vicinity of sewer outlets or near outlets of large streams. Therefore, minimum treatment is required. The availability of practically unlimited quantities of water is also a decided advantage. Unfortunately, however, the sewage from a city is often discharged into the same lake from which the water supply is taken. Great care must be taken in locating both the water intake and the sewer outlet so that the pollution handled by the water treatment plant is a minimal.

Sometimes the distance from the shore where dependable, satisfactory water can be found is so great that the cost of water intake facilities is prohibitive for a small municipality. In such cases, another supply must be found, or water must be obtained from a neighboring large city. Lake water is usually uniform in quality from day to day and does not vary in temperature as much as water from a river or small impounding reservoir.

Impounding Reservoirs. Impounding reservoirs have the advantage that most of the turbidity may be removed by settling out during storage. There may be additional benefits, such as a reduction in bacteria. But there may be disadvantages, such as the production of unpleasant tastes and odors by algae. This is discussed along with other factors of self-purification in a later chapter.

Sanitation and Control of Water Sources. It is always more desirable to prevent the pollution of sources of water supply than to rely entirely upon the effectiveness of the water treatment process. Rules and regulations specify conditions for the construction and maintenance of leaching cesspools, privies, and sewer outlets. Other sources of pollution in the reservoir drainage also are regulated.

Erosion control in the drainage area is of value in reducing the rate of silt deposition in the reservoir and the amount of turbidity which the treatment plant must remove. It is desirable for a municipality to own the entire drainage area and to follow proper land use practices within its boundaries. If funds for the purchase and reforestation of the area are limited, the area immediately surrounding the reservoir should be reforested and extended when additional funds become available. Competent advice should be secured on the type of plantings to be used in the reforestation of watersheds.

Catchment areas occasionally contain swamps which discolor the water. Shallow pools of water in a swamp area frequently support a prolific growth of microorganisms which may be washed into the outlet stream. Many swamps may be drained by ditching and by cleaning existing channels. Such improvements are very desirable because the reduction in the color and the content of microorganisms will be permanent if proper supervision and maintenance are provided.

Generally, groundwater supplies are used only by smaller communities because of the limited quantity that can be obtained from an aquifer. An objection to an underground supply is that the water may be excessively hard. This is due to the percolation of the water through mineral deposits from which constituents which cause hardness are leached. On the other hand, an underground supply usually has the advantage of requiring a minimum degree of treatment unless polluted because of natural purification as the water passes through the various underground soil formations. These conditions are general. Some mineral deposits do not contribute to hardness and some underground formations do not effectively remove objectionable material. The quality of such water supplies has come under increasing scrutiny and regulation because of pollution.

Types of Groundwater Sources.

Subsurface water is schematically shown in Figure 1-4. Groundwater sources are most often wells. Infiltration galleries and springs are sometimes used. Wells may be classified as to the type of aquifer from which they derive the water or as to the type of construction. If it draws water from an unconfined aquifer, it is called a free surface well, gravity well, or water table well. An unconfined aquifer is one which contains water with a free surface at atmospheric pressure. If the well draws water from an aquifer which is confined both above and below, and

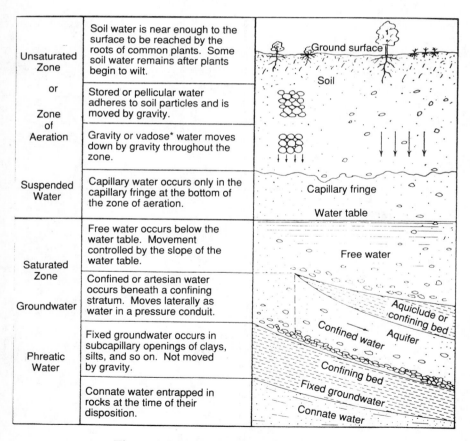

Figure 1-4 Occurrence of subsurface water

contains water under pressure, it is called an artisan well and the aquifer is known as an artisan aquifer. Free surface wells tend to be shallow and the artisan wells tend to be deep. Shallow wells are commonly dug or driven wells, and deep wells are usually drilled (see Figures 1-5 and 1-6).

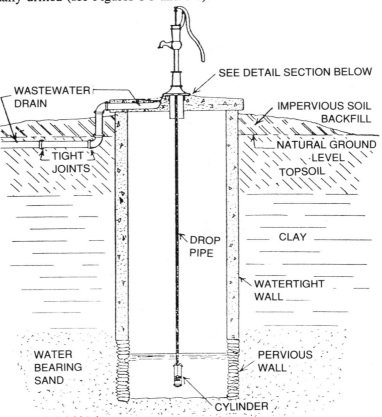

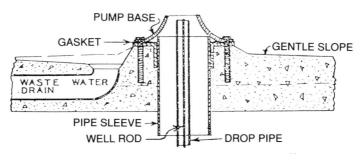

Figure 1-5 Typical shallow dug well

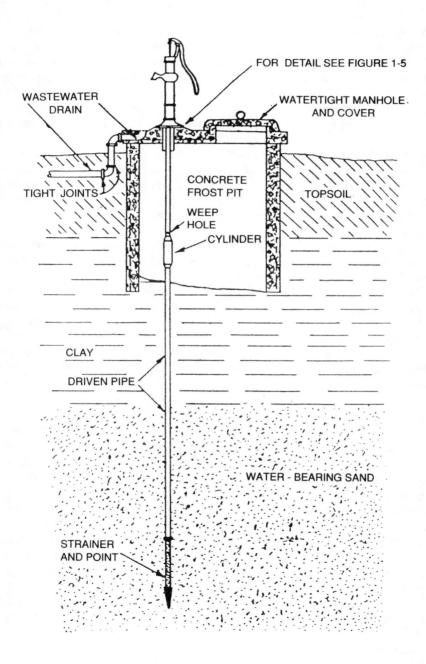

Figure 1-6 Typical shallow driven well

A dug well consists of a vertical hole, usually 4 to 6 feet in diameter, which is excavated through topsoil until the water bearing stratum is encountered. Such wells may be lined with concrete, brick, rough stone, or vitrified tile. That portion of the lining extending from about 1 foot above the ground surface to at least 10 feet below the surface should be watertight to exclude any surface drainage. Concrete is the best material for the upper part of the well; brick, rough stone, or vitrified tile is usually used for the pervious section which extends down into the water-bearing stratum.

Driven wells may be used when the groundwater is within about 25 feet of the ground surface and when there are no boulders or intervening rock formations. Such wells are easily constructed and can be protected against superficial pollution, although they are more subject to pollution than deep wells penetrating impervious material. The simplest type of driven well consists of a perforated brass strainer which is one shaped at the end and attached to the bottom of an iron pipe. The pipe is driven through the upper soil layers and into the water-bearing material.

Deep Wells. When the topsoil strata are too impervious or do not contain groundwater, wells must be either drilled into rock to obtain water from crevices or through the soil to deeper water-bearing strata (aquifers). Although the elimination of nearby sources of pollution is important, it should be remembered that polluting material may pass through fissured rock for great distances without the improvement in the quality of water which would result from filtration through the subsoil.

Drilled wells are 4 to 12 inches in diameter, although in some instances they may be larger. Metal casings can provide effective protection against the entrance of surface water and polluted groundwater, if the casing is sealed to prevent the entrance of all sources of pollution. When the water-bearing material is sand or gravel, a well screen is usually attached to the bottom of the casing. Since water can only be lifted a maximum of about 20 feet by suction, the pump must be located below the ground surface near, or below, the water level in the well.

Gravel-packed wells are often used when the water-bearing stratum contains fine sand. Such a well is similar to a standard well except that for several inches around the well screen the sand has been removed and replaced with gravel. This type of construction increases the area of contact with the water-bearing stratum. Resistance to the flow of water into the well area is decreased and the capacity is increased. The gravel outside the screen also helps to prevent sand from being drawn into the well during pumping. The overall operating results of a gravel-packed well have been very satisfactory and some authorities advocate gravel packing as standard practice when tapping any water-bearing stratum of unconsolidated material. The actual placing of the gravel may be done in a number of ways but requires skill and experience on the part of the well driller. Figure 1-6 shows typical deep well developments.

Springs. Springs occur where a water-bearing stratum reaches the surface of the ground or where fissures in rock "outcrop" at the surface under conditions so that the groundwater is forced through the fissures. The first type of spring is usually of local origin and great care must be taken to eliminate nearby sources of pollution. The origin of the water flowing from a rock spring is difficult to ascertain without detailed knowledge of the geological formations in the area.

Many methods have been developed to collect spring water. When the water does not emerge from the ground at a well defined point, it may be collected and carried to a well or basin by tile lines with open joints laid in ditches at right angles to the direction of underground flow. The tile is surrounded with crushed stone gravel and the ditch is filled with clay to keep out any surface drainage (see Figure 1-7).

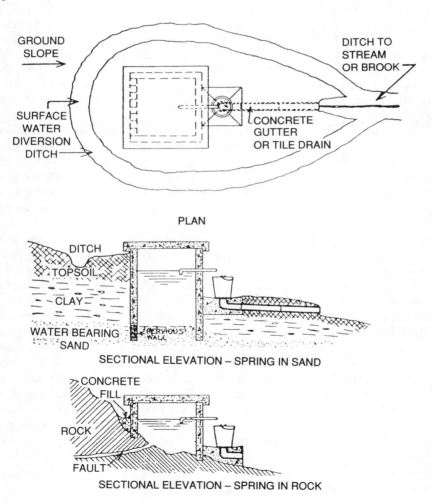

Figure 1-7 **Typical spring basin construction.**

Regardless of the type of construction, all spring basins should be covered, and the water source should be permanently protected by a structure of concrete or other impervious material so all water except that issuing from the spring will be excluded.

Surplus water should be piped from the covering structure in such a manner that surface water cannot enter the spring even during floods. It is not necessary to ventilate spring structures; all openings should be covered. Inspection manholes should be fitted with tight locked covers.

Groundwaters. Waters obtained from springs and wells in a given locality may be very different from the local surface waters. Most well and spring waters indicate the nature of a local geologic area, while running streams reveal the conditions of their drainage areas and thus provide an approximate average index of the geology of a much larger territory. Stream waters are usually the lower in dissolved mineral content but vary greatly with the change of seasons in dissolved impurities, turbidity, and chemical character. Groundwaters are more nearly constant in composition but are likely to contain deleterious gases, relatively high concentration of mineral constituents, and are the more pronounced in chemical type.

Groundwaters are normally more concentrated as the level from which they are drawn deepens. Older crystalline rocks afford waters of low concentration free from permanent hardness, although temporary hardness and a relatively high content of silica may be expected. Such waters are prevalent in the upper Mississippi River basin, in the region from the Appalachian Mountains to the Atlantic coast, from New England to the Gulf of Mexico, and in relatively small areas in various sections of the United States. Waters from basalts and rhyolites, such as are found in many parts of the Rocky Mountain and Pacific Coast states, are likely to be somewhat more concentrated and may carry more sulfate. Sedimentary rocks, particularly marine sediments, such as are found throughout the greater part of the Mississippi Valley, afford water of still greater concentration. Waters from limestones are very hard and those from magnesia limestones and dolmetic limestones carry both scale-forming and corrosive ingredients.

Sandstones differ greatly in the content of soluble constituents and in the quality of waters, which is true also of shales. The best and worst of industrial waters may come from shales of widely differing character. Many sandstones and shales afford waters of high permanent hardness. These rocks, like the limestones, cover great areas, and once the character of the water from any particular stratum is determined, the general characteristics of a water from the same stratum many miles distant can be forecast with a fair degree of accuracy. In general, groundwaters are not as desirable for industrial purposes, but may be as satisfactory for sanitary use as surface waters. This may be inferred from their tendency to higher mineral content and lower turbidity and organic content.

Groundwater Biology

Groundwater is a dramatic contrast to surface streams as a life habitat. Fluctuations of terrestrial climate are affected by increasing the distance from the earth's surface, and at still greater depth the effects of geophysical processes may predominate. Chemical composition of groundwater may change not only by inorganic processes but also through the activity of microorganisms. Deep-seated groundwaters may contain a specific microflora which actively influences their chemical composition.

Bacteria are the significant forms of life in groundwaters. Such microorganisms are not uniformly distributed horizontally or vertically but depend on the distribution of organic matter, mineral species, sediment pore and particle size, and the intensity of water exchange. There is also an optimum temperature which favors bacterial activity, and the depth at which the optimum occurs depends upon the local geothermal gradient. The increase in geothermal heat with increasing depth eventually precludes bacterial life at great depths.

Many of the most important chemical effects are caused by microorganisms whose life activities do not require the presence of preexisting organic matter. Chemosynthetic, autotrophic microorganisms grow and reproduce by using inorganic food and the energy from the oxidation of inorganic compounds such as molecular hydrogen, methane, and reduced compounds of nitrogen, sulfur, and iron. Of particular interest are the anaerobic organisms, which are able to grow using both organic and mineral substances in the absence of oxygen. Most available literature discusses biogenic transformations of sulfur, nitrogen, carbon, iron, and manganese, but the details of the inorganic and organic processes involved are very imperfectly known.

Varying events may occur with microorganisms as groundwaters move in sediments. Active or passive migration of the cells with or without multiplication may take place within the streaming water and the sediment; cells may grow through the sedimentary material, from surface films, attach to the surface of particles of the sediment, or be retained, adsorbed, or eluded within the sediment.

Besides the physical laws which regulate retention and elution, biological differences are found among the different bacterial species. This is a very brief review of groundwater biology which includes only the most general aspects of the ecosystem. However, the ecology of groundwaters is so poorly understood that even conceptual models of biological effects on water quality probably cannot be developed from existing data. Refer to Chapter 6 for further discussions on groundwater.

Water Consumption

The rate of use of water in public systems varies greatly and depends upon the size and character of the community. In most public systems in the United States it can vary between 60 and 300 gallons per capita per day. About 45 percent of the total is for domestic use and an equal amount is used for industrial and commercial purposes. The remaining 10 percent is for public and other uses. The estimated amount of leakage in distribution systems varies from about 5 percent to over 40 percent of the water delivered, and averages about 10 percent to 15 percent.

Uses of water are usually classified as domestic, public, commercial, and industrial. Domestic use includes water utilized in residential units for drinking, cooking, bathing, washing, heating, cooling, air conditioning, sanitary purposes and watering lawns and gardens. It is population dependent and accounts for 30-50 percent of the total water consumption.

Public use of water is for places and buildings such as public gardens, parks, and fountains, public swimming pools, hospitals, schools and other educational institutions, prisons, public sanitary places, street and sewer flushing, and fire fighting and amounts to 5-10 percent of total consumption.

Commercial use of water includes sanitary and air-conditioning purposes, in office buildings, hotels, and restaurants, car washing, laundries, golf courses, shopping centers, bus, railway and air terminals, and so on, and can amount to 10-30 percent of the total consumption.

Industrial use for manufacturing and processing consists of heat exchange, cooling, and manufacturing and bears no relation to the population of an industrial area and is expressed in terms of the unit of production. It accounts for 20-50 percent of the total consumption.

In addition to the preceding direct uses, 10-20 percent of the total consumption may be due to leakage through pipes and valves, evaporation from reservoirs, and so on.

Factors Affecting Demand

Water demand for a community is typically expressed as the average consumption per capita per day and is computed by dividing the total consumption in a year by the population and the number of days in the year. Average consumption depends upon several factors and increases with:

- the increase in the size of the community due to increase in the demand
- improvement of economic status of the community
- improvement in the quality of the water
- increase in pressure in the distribution system
- provision of sewer facilities

Average water consumption decreases with:

- increase in the cost of water
- the use of meters

Average consumption also depends upon the climate. In hot and arid climates, the average consumption can increase due to frequent bathing, air cooling or air conditioning, and heavy lawn sprinkling. In extremely cold climates, the average consumption increases due to the bleeding of water to avoid freezing in the service pipes and the internal piping systems. Furthermore, the type of supply such as continuous 24 hour or intermittent water supply is for extremely short duration; the average consumption is more for the intermittent supply than for continuous supply. In the intermittent supply, water that is stored but remains unused during no-supply periods may be discarded when a fresh supply begins. Water demand for fire-fighting purposes depends on the population of the community, types of industries, and the duration of fire. The fire demand established by the American Insurance Association is expressed as

$$Q = 3,865 \sqrt{P}(1 - 0.001 \sqrt{P})$$

where Q = rate of fire demand in L/min, and P = population in thousands.

Water demand fluctuates with the seasons of the year, the days of the week, and the hours of the day. Fluctuations are greater as the size of the community and the duration of time decrease. The estimated amount of leakage in distribution systems varies from about 5 percent to over 40 percent of the water delivered, and averages about 10 percent to 15 percent. In any system, the consumption rate varies continuously and the range of variation depends upon the size and character of the community served.

Water Supply Requirements

The quantity of water required daily in a district is usually stated in terms of per capita consumption, and the figure includes domestic, commercial, and public use. Domestic demand of 40 gallons per capita is considered average. Commercial use varies greatly according to the type of community and represents a consumption comparable with domestic use. Public use and leakage together represent also a similar quantity. Thus, the total water requirements for all purposes of the average will be somewhat above 100 gallons per day on the per capita basis. Exceptions are found in industrial centers and depend on the type of industrial activity.

Such figures do not include the consumption of industrial plants with private water supplies. In many of these, the principal consumption is for cooling or condensing purposes. The purity requirements of water supply vary according to the specific industries served and these must be considered individually.

WATER DISTRIBUTION

Water of the desired quality is distributed to the consumers in sufficient quantity and at adequate pressure through a water distribution system. This includes pipes, valves, hydrants, booster pumps, reservoirs for distribution and equalizing purposes, service pipes to the consumers and meters, and all other appurtenances after the water leaves the main pumping station or the main reservoir.

The layout of a distribution system should follow the street pattern with interconnected mains providing loops. The loops provide alternate routes in the event of pipe failure or pipe closure and, therefore, they increase the reliability of water distribution. Dead ends, when provided, should be as few as possible as they cause stagnation of water and water quality problems. They should be periodically flushed. A branching or a tree system is cheaper, even though less reliable as compared to a looped system and, therefore, is used for water supply to small communities.

For equalization of water supply in a large water distribution system, zoning in the distribution system must be provided. Zoning depends upon the density of population, type of locality, topography, and the facility of isolation for waste assessment and leak detection. A separate system should be provided for each zone having an average elevation difference of 40-50 meters. Neighboring zones are interconnected to provide emergency supplies; but the valves in the interconnecting pipes should normally be kept closed.

Minimum pipe sizes of 150 mm in residential areas and 200 mm in high-value districts should be provided. However, 100 mm pipe sizes can be used for short lengths in residential areas and for distribution systems of small communities. The maximum length between cross connections should be limited to 180 m. Valves should be so located that in case of pipe breakage, the pipe length required to be shut off should not exceed 150 m in high-value districts and 240 m in other areas.

Distribution and Equalization

Reservoirs are provided in distribution systems to ensure sufficient quantity and pressure. These are connected to the distribution system by separate inlet and outlet pipes, or by a single pipe such that water enters and leaves by the same pipe. In the latter case, the reservoir is said to be "floating on the system." When the supply rate exceeds the demand rate, water enters into the reservoir; when the demand rate exceeds the supply rate, water flows out of the reservoir.

Reservoirs can be surface reservoirs or elevated reservoirs. Surface reservoirs are used when topography permits. They are square, rectangular or circular in shape and are built in earth, masonry, concrete, or reinforced concrete. The depths vary from 3 m to 10 m, usually increasing with the capacity. When the height of a surface reservoir exceeds its diameter, it is called a *standpipe*. It is built in steel or reinforced concrete; however, steel is more preferable because of its watertightness. Elevated reservoirs or tanks of difficult sizes and shapes in steel, reinforced concrete, and prestressed concrete are used. Altitude valves which automatically shut when water reaches a predetermined level are provided. An overflow with drainage capacity equal to the maximum rate of inflow is provided for safety against the failure of the altitude valve. Water-level indicators, preferably with remote-recording equipment, should be provided on the reservoirs to know their performance.

Reservoirs which mainly provide a sufficient quantity of water are termed *service, storage,* or *distribution* reservoirs; while the reservoirs which mainly ensure adequate pressure are termed *balancing* or *equalizing* reservoirs. Distribution reservoirs serve as balancing reservoirs and balancing reservoirs serve as distribution reservoirs.

Distribution reservoirs provide:

- Storage for equalizing supply and demand over periods of fluctuating consumption in a day. The reservoir fills when the rate of filtration or pumping exceeds the demand rate and empties in reverse conditions. This allows the pumps and the treatment plants to operate at a constant rate and thereby to reduce their capacity. The required storage is obtained for hourly demand fluctuations on a maximum day and for the proposed period of filtration or pumping. This is computed in a tabular form or determined graphically from either the mass curves as explained earlier, or from the maximum area between the supply-rate and demand-rate curves.
- Fire storage for immediate use in large quantities for fire-fighting purposes. The required fire storage depends upon the population served.
- Emergency storage for provision during failure of intake, supply conduit, or power. The required emergency storage depends upon the extent of damage and the corresponding time of repair, and also the capacity of the supply conduit.

In a distribution system with a single distribution reservoir, its ideal location is a central place having economical pipe sizes and equitable pressures. When the system is fed through direct pumping and also through a distribution reservoir, its location should be at the tail end of the system. The capacity of the distribution storage is obtained by reasonably combining the preceding three purposes. If a major fire occurs on a day of maximum demand, combined capacity for these two purposes should be provided. The simultaneous requirement of emergency storage is less likely and it would also require a much larger capacity and therefore is not considered. However, emergency storage is considered separately, and a reasonable choice of the capacity for the distribution reservoir is made.

Equalizing reservoirs are used for:

- Equalizing pressures in the distribution system and thereby reducing fluctuations in pressure caused by fluctuating demand. This provides better service to consumers and pressures for fire hydrants.
- Raising pressures at points remote from distribution reservoirs and pumping stations and improved service during periods of peak demand.

- Equalizing head on pumps. When equalizing reservoirs are located near pumping stations, the pumping heads are more uniform. This results in better selection of pumps and their operation at higher efficiency.

Equalizing reservoirs are usually built at the opposite end of the distribution system from the source of supply. They should have sufficient elevation to provide adequate pressure in the system. The required capacity is obtained from the mass curves or the supply-rate and demand-rate curves.

The analysis of a water distribution system involves the determination of the unknown parameters with the help of known parameters. The length, diameter, and roughness characteristics of a pipe are usually known and therefore are grouped into one parameter—the pipe resistance constant. Therefore the head loss h_p in a pipe p can be expressed by a general relationship

$$h_p = H_i - H_j = R_p Q_p^n$$

in which H_i, H_j = hydraulic gradient level values at the end nodes i and j, respectively, of the pipe; R_p = resistance constant of pipe p; Q_p = discharge in pipe p; and n = an exponent which is 1.85 in the Hazen-Williams relationship and 2.0 in the Darcy-Weisbach relationship.

Water is withdrawn along the lengths of the pipes through house connections. However, for simplicity, it is common to assume in the analysis as well as in the design that the withdrawals are concentrated at their ends. Thus, all the withdrawals are assumed to be concentrated at pipe junctions or nodes. These nodal withdrawals or nodal demands are determined by considering the area served by each node, density of population, per capita consumption, peak demand factor; and specific commercial, public, industrial, and fire demands. Even though the nodal demands fluctuate with time, it is common to assume the demand pattern and the flow in the distribution system to be steady.

For distribution systems, the involved parameters are the pipe resistance constants, the nodal flows (inflows at supply nodes, that is, reservoirs and pumps, and outflows at demand nodes), the pipe discharges, and the hydraulic gradient level values at the nodes. Some of these parameters are known and the remaining parameters are determined in the analysis. Any combination of known or unknown parameters can be provided of the combination satisfies certain conditions. The analysis of the distribution system usually involves the determination of the inflows at the supply nodes and the hydraulic gradient-level values (or the pressures) at the demand nodes.

Quality

Water supplied to consumers should be free from pathogenic organisms, clear, palatable, free from undesirable taste and odor, of reasonable temperature, neither corrosive nor scale forming, and free from minerals which could produce undesirable physiological effects. It is desirable to protect the water sources from being contaminated by pollution. When such protection is difficult or impossible, water must be treated and brought to a level of an acceptable standard. Proper care should also be taken during the transportation and distribution of treated water to ensure that it does not get polluted through joints or backflow of polluted water from customer piping.

To ensure the quality of water, certain standards have been developed by the U.S. Public Health Service, World Health Organization, and others. Table 1-2 gives the physical, chemical, and bacteriological standards for maximum acceptable and maximum allowable concentrations.

Hardness in water is predominantly caused by carbonates and bicarbonates of calcium and magnesium, and sulphates and chlorides of divalent cations. Iron in water causes taste and hardness, discoloration of clothes and plumbing fixtures, and incrustation in water mains. Manganese has an effect similar to iron. Manganese is also oxidized into a sediment which clogs pipes, discolors fabrics, and stimulates organic growths. Sodium chloride adds salinity to water. Fluorides, when more than 3.0 mg/l (even more than 1.5 mg/l), result in mottled and discolored teeth in children of formative age. Fluorides also reduce the strength of bones.

SOURCES

Major sources of water for water supply purposes are surface water and groundwater. Surface water includes fresh water available in natural lakes, rivers and streams. Groundwater includes water obtained from wells, springs, and infiltration galleries. Where these sources are scarce in comparison with the demand, attempts are made to augment water supply through desalination, wastewater reclamation, artificial precipitation, and also through reduction of evaporation losses in reservoirs.

Factors influencing the selection of source for water supply include:

- Quantity of water available should be adequate to meet the requirements.
- Water should be of good quality, that is, free from toxic, poisonous, and other ingredients injurious to health. The impurities should be as few as possible and easily removed.
- Reliability of a source depends upon the discharge rate and the elevation. The decreasing order of reliability is: discharge rate adequate and also at sufficient elevation so that supply is feasible without storage and pumping; enough quantity feasible with storage at sufficient elevation so that supply is possible through storage and under gravity only; rate of discharge sufficient but at an adequate elevation so that storage is not necessary but pumping is required; and source that requires both storage and pumping.
- Source should be as close to the demand as feasible.

Table 1-2 Water Quality Standards

Substance or Characteristic	Maximum Acceptable Concentration	Maximum Allowable Concentration
Color on platinum cobalt scale	5	50
Taste	Unobjectable	—
Odor	Unobjectable	—
pH range	7.0 to 8.5	6.5 to 9.2
Total dissolved solids	500 mg/L	1,500 mg/L
Turbidity in NTU (Nephalometer Turbidity Units)	5	25
Total hardness (as $CaCO_3$)	200 mg/L	600 mg/L
Calcium (as Ca)	75 mg/L	200 mg/L
Chloride (as Cl)	200 mg/L	600 mg/L
Copper (as Cu)	0.05 mg/L	1.5 mg/L
Fluorides (as F)	1.0 mg/L	1.5 mg/L
Iron (as Fe)	0.1 mg/L	1.0 mg/L
Magnesium (as Mg)	30 mg/L	150 mg/L
Manganese (as Mn)	0.05 mg/L	0.5 mg/L
Nitrate (as NO_3)	45 mg/L	45 mg/L
Sulphate (as SO_4)	200 mg/L	400 mg/L
Zinc (as Zn)	5.0 mg/L	15 mg/L
Anionic detergents (as MBAS)	0.2 mg/L	1.0 mg/L
Mineral Oil	0.01 mg/L	0.30 mg/L
Phenolic compounds (as phenol)	0.001 mg/L	0.002 mg/L
Toxic substances		
Arsenic (as As)	—	0.05 mg/L
Barium (as Ba)	—	1.0 mg/L
Cadmium (as Cd)	—	0.01 mg/L
Chromium (as hexavalent Cr)	—	0.05 mg/L
Cyanide (as CN)	—	0.05 mg/L
Lead (as Pb)	—	0.1 mg/L
Mercury (total as Hg)	—	0.001 mg/L
Selenium (as Se)	—	0.01 mg/L
Polynuclear aromatic hydrocarbons (PAH)	—	0.2 µg/L
Radioactivity levels		
Gross alpha activity	—	3 pCi/L [a]
Gross beta activity	—	30 pCi/L
Bacteriological levels		
E Coli count in any 100-mL sample	—	0
Coliform organisms in 95% of 100-mL samples in any year	—	0
Coliform organisms in any 100-mL sample	—	10
Coliform organisms in second 100-mL sample after it is found in first sample	—	0

[a] pCi = pico (10^{-12}) curie

- Legal aspects are important in groundwater sources and also in transfer of water from one watershed to another. Political aspects creep in through interstate and international river water disputes and agreements for surface waters.
- Cost is an important factor and should be as low as possible.

Water Quality and Source

Surface waters may contain silts and suspended sediments, dissolved organic matter from topsoil, bacteria and disease germs from sewage, chemical impurities from industrial wastes, and fertilizers and toxic materials from pesticides.

Waters from rivers and streams carry a considerable amount of silt and suspended sediment and are colored during floods but are comparatively free from turbidity and color during other seasons. River waters may be polluted from sewage and industrial wastes. Lake waters, even though free from turbidity, are subject to pollution.

Surface waters when stored in impounding reservoirs become clearer due to settling of the suspended sediment and also become colorless due to coagulation and settlement of the suspended matter and the bleaching action of the sun's rays. Storage also reduces bacteria and pathogenic organisms; however, the quality of water in impounding reservoirs might deteriorate due to bacterial decomposition of the reservoir floor and the solution of iron, manganese and hardness-producing calcium and magnesium minerals. Further, water from the lower strata of deep lakes and reservoirs contains less oxygen and more carbon dioxide as compared to water from surface layers.

Groundwaters are free from silt and suspended load and are also free from sewage and industrial wastes if such pollution sources are not nearby. However, groundwaters may contain dissolved impurities like salts of calcium, magnesium, iron, manganese, sodium, fluorine, and others collected from soils and rocks through which they percolate. Deep well waters are not affected by rainfall and therefore their quality does not seasonally change. They are also relatively free from harmful organisms but are hard. Shallow well waters may be relatively soft but are more affected by rainfall and pollution than deep well waters. Other considerations for groundwater are discussed in Chapter 6.

Reservoirs

Impounding reservoirs store water during high periods of runoff and release it for use during low periods of runoff. Sites for impounding reservoirs depend on the quality and the quantity of water available, geological considerations of the reservoir and dam site, the distance from and the elevation relative to the demand center, and economic considerations.

The amount of storage provided in an impounding reservoir depends upon the height to which a dam can be built, the storage required to satisfy the demand,

and the net supply, that is, the safe yield available from the watershed (safe yield is the maximum net volume available from the watershed after deducting the losses due to evaporation from the reservoir surface; infiltration, percolation, and leakage through the reservoir floor and dam; and withdrawal for downstream riparian owners).

The required capacity of the impounding reservoir can be obtained from the supply and demand mass curves. If the demand rate is constant, the demand mass curve is a straight line.

Municipal Supply and Other Requirements

The quality requirements of a municipal water supply are concerned primarily with physical and biological standards and to a minor extent with chemical composition. Such water supply should be derived from a source free and protected from pollution and the mean density of bacteria of the bacteria of the B. coli type set by regulatory standards. It shall be clear, colorless, odorless, and pleasant to the taste and shall not contain an excessive amount of soluble substances or any chemical employed in treatment.

The upper limits expressed in parts per million for certain soluble substances are set for iron, magnesium, sulfate, chloride, and total solids. Municipal water supplies must meet these limits and a water just within the limits would be highly unsuited to most industrial purposes.

To encourage the establishment of industries within their areas, many municipalities, provide a water supply which will meet not only sanitary standards, but which is also suited to specific industrial requirements. To accomplish this, they may spare no expense to bring high-grade waters from a considerable distance and often install extensive purification plants for the removal of undesirable impurities.

Irrigation Waters. It might seem that water for this purpose need not be different from ordinary river water, but irrigating water for the most part is eventually evaporated from the soil over which it is spread and therefore leaves all its soluble impurities behind. These gradually concentrate in the surface of the ground and may eventually become detrimental to plant growth (See Figure 1-3). Land would probably be injured by the best of natural waters if irrigated with them for too long a period of time without natural or artificial drainage. High-alkali content is particularly detrimental and the value of a water for irrigating purposes is roughly established by its content of alkaline substances. The higher the alkalinity the greater is the portion of the water which will have to be drained away from the irrigated area to keep it flushed out. The value is decreased not only because of the poorer quality but because of the larger quantity required.

Textile Industry. Textile fibers including wool, silk, and cotton are prepared for fabrication by scouring processes which involve the use of quantities of water. The standards of purity required are high in all cases but vary with the specific operations.

In wool scouring, all fresh water is added in the third or final scouring tank from which the wool emerges in its finished condition. Impurities in the water that will cling to the fiber, therefore, are obviously out of place and will have the same deleterious effects in subsequent processes as imperfect scouring. Suspended impurities are generally undesirable, organic matter being especially deleterious. Therefore, waters polluted by sewage or organic industrial wastes and turbid waters are always purified before use in scouring tanks.

A more important class of impurities consists of those substances which unite with soaps used in scouring to form insoluble soaps. These substances, iron, aluminum, calcium, and magnesium, destroy their equivalent values of soap for detergent purposes, thereby making necessary the use of an excessive amount of soap. In addition, the insoluble soaps stick closely to the fiber and make efficient scouring very difficult. On this account, they are more detrimental to the process than inorganic suspended matter composed largely of colloidal particles of clay.

In scouring by the two-stage process of steeping and scouring, the importance of pure water is even greater because in addition to the effect of hardness in the scouring operation, the recovery of potash from the steep water will be rendered somewhat more expensive and the purity of the product will be decreased by the saline constituents of the water. In this case all mineral impurities, not merely the hardening constituents, have a deleterious effect. In scouring by volatile solvents and subsequent washing, since no soap is used, the purity of the water is a minor consideration. Suspended matter is undesirable. If the wash water is utilized for the recovery of potash, all dissolved impurities will assume importance inasmuch as some will cause scale in the evaporators and all will decrease the purity of the potash produced. Each 100 parts per million of mineral solids in the water will reduce the purity of the potash by approximately 1 percent. If the wool is scoured by detergents such as sulfonated alcohol, then the hardness in the water will be of less consequence. This is due to the fact that the calcium and magnesium compounds of these detergents are soluble; hence, they do not cause a loss of detergent power and more particularly, do not precipitate on, and foul the surface of the wool. Water used for rinsing the wool after scouring may amount to as much as 100 gallons for each pound of wool scoured. This water should be free from suspended matter, but otherwise requires no special qualities.

The cotton industry has long recognized the value of soft water in its operations. The boiling-off has required the use of soap in large amounts. Such soap, like others, is wastefully consumed by calcium and magnesium in the water. The precipitates formed are calcium and magnesium resinates which are just as troublesome in later operations as the corresponding oleates and stearates. Spots which are difficult to wash out, dye, and bleach are formed and a uniform white, so essential in this industry, is almost impossible to obtain. It is characteristic of silks and cottons produced by plants using soft water that the fabric has a softer feel, not at all like the harsh feel of fabrics produced with waters containing large amounts of the hardening constituents. In the manufacture of textiles, therefore, the necessity for a clear, colorless water containing as little calcium and magnesium as possible is obvious unless the newer detergents are used.

In the rayon industry as in the case of natural fibers the water must be of high purity to give best results. Water must be soft because lime and magnesium salts which are responsible for the hardness of water also cause hardness of finish on rayon and interfere with level dyeing. Traces of iron in the water may be highly detrimental. Splotchy discolorations in the product are possible.

Water for Dyeing and Bleaching. Because using water for dyeing and bleaching involves such small amounts of chemicals and large amounts of water to produce delicate coloring effects, the dyeing of fabrics often presents chemical problems of considerable difficulty. The solution of these problems often hinges on the proper choice or treatment of the water used in the process. A water containing large amounts of calcium is unsuitable and even ruinous for use in dyeing with aniline colors but is essential to the successful use of logwood or weld, dyed on a mordant of iron or aluminum. Calcium and magnesium act very much alike in dyeing. Some of the effects of these constituents are as follows: Heavy tarlike precipitates form when such aniline colors as methyl violet, malachite green, magenta, and safranine are dissolved in waters high in calcium and magnesium. This tarlike precipitate sticks to the fiber and results in uneven dyeing, poor shades, and that annoying defect of "rubbing off." Magenta and safranine are peculiarly susceptible to the influence of these constituents, the effect being noticeable in changes in both color and intensity of the coloring material. Color is wasted and flat shades are produced in dyeing with Turkey red or cochineal scarlet. On the other hand, in dyeing with alizarin it is necessary to have calcium carbonate present to cause complete saturation of the mordants, but when calcium and bicarbonate radicals are present, carbonic acid is freed, calcium carbonate is precipitated by the heat, and the bath takes a violet color due to the formation of a compound of calcium and alizarin. If the solution is boiled, the lime lake deposits as a violet powder, and the bath cannot be used. Thus, it can be seen that calcium may be helpful or detrimental to dyeing according to circumstances, but as a rule, calcium and magnesium salts result in uneven dyeing, fading colors, spotted effects, off-shades, and a waste of color.

Iron is a very objectionable impurity. Dull and flat colors are frequently caused by this constituent, especially when dyeing on a mordant. If water contains iron in any appreciable amounts, it is practically useless for dyeing. Also in bleaching, iron is troublesome. The iron is oxidized by the bleach solutions, causing yellow, brown, or muddy white effects. This difficulty can be partially remedied by acid treatment, but there is then danger of injuring the fabric. Laundries have the same difficulty with ferruginous waters. It is therefore evident that a water suitable for dyeing and bleaching should be free from suspended matter, low in its content of iron, and as soft as possible.

Water in Paper Making—While the ordinary constituents of a water which contains permanent and temporary hardness cause little difficulty in preparing the pulp, their effect in sizing is harmful. Sizing in precipitating, with alum (aluminum sulfate), and the resin from a solution of a resin soap. The aluminum resinate thus formed is precipitated on the fibers. Reactions between lime and a soap here take place if the water is of a calcareous nature. Carbonic acid decomposes the resin soap, forming sodium bicarbonate and, if calcium is present in appreciable amounts, calcium resinate. This compound has a more granular consistency than the corresponding aluminum compound, does not adhere to the fibers as well, and hence is entirely unsuited for sizing purposes. Alkaline waters give trouble through the formation of double salts with alum, causing incomplete or improper precipitation of the aluminum resinate. The difficulty here may be overcome, of course, by the previous treatment of the water with alum, but unless this is a part of a purifying system to which careful attention is paid, large amounts of alum are wasted with but partial prevention of the trouble and variable results.

Turbid waters, or waters carrying suspended matter, are injurious to the product. Such particles resist the action of bleach and cause dark specks and spots in the finished paper, particularly noticeable in white and light shades of fine writing and art papers. Vegetable coloring matter produces either streaks or a dull shade imparted to papers that should be white. Acid waters are among the most annoying offenders. They not only give trouble with the size, decompose the colors, and produce streaks in the papers, but they also attack the expensive wire screens and dryer-felt of the machine.

In making white or light shades of paper, ferruginous waters are out of the question unless the iron is removed. Alkali precipitates ferric hydroxide, which gives the pulp a brown color. During sizing the iron also gives trouble, and rust spots in the finished paper are a common result of the use of water containing iron.

Tanning is an industry in which the quality of the water has played an important part. The large volumes of water used both in the dehairing and in the tall pit make necessary the careful selection or purification of the water supply. Dehairing consists in loosening the roots of the hair by treatment of the hides with quicklime. If the water used in this process contains bicarbonates or free carbonic acid, there results a precipitation of calcium carbonate in the dermic tissues. This precipitate naturally interferes with the absorption by the cells of the hide, of the tanning which is intended to convert the skin into the insoluble material, leather. In the tan pits, calcium and magnesium react with the tanning, destroying its usefulness. A considerable amount of an expensive chemical is thus wasted and, according to one authority, the secondary products formed impart to the leather a reddish brown color which lowers the market value of the finished product. One way of obviating this difficulty is to treat the water with a free mineral acid, preferably sulfuric. The acid seems to aid in swelling the hides and if used at the proper concentration for a period of time not too great, no rotting of the hides takes place. Chlorides are harmful to the proper swelling operation since they do not unite with the hide and even prevent the proper action of the acids, but sulfates of calcium and preferably are desirable. In their presence plumping is more readily effected, the finished leather has a finer and more compact grain, and the cut surfaces are cleaner and shinier.

Organic matter should be absent from water used in tanneries, since one of the prime requisites of the finished product is that it must not decay. Organic matter tends to cause decay or "leather rot." Hence, its removal becomes a matter of necessity rather than choice.

Iron is objectionable in the water supply of tanneries, since the same reactions which are used in making one kind of black ink take place between the iron of the water and the tannin with which the hides are treated. In both cases, ferrous iron is oxidized by the air to ferric, which then reacts to form black tannate of iron. Such coloration is, of course, ruinous to any but black leathers. Water for tanneries should therefore be clear and low in iron, hardness, and chlorides though calcium and magnesium sulfates are harmless for certain of the operations.

Other chemical industries of importance depend to some extent on the quality of their water supply for successful maintenance. The brewing industry is one requiring large amounts of water both in the malting and fermenting process and it is recognized that mineral salts may be very injurious in both of these processes. In the case of malting, a low concentration of sulfate is not altogether undesirable but, if carbonates are present in either this or the later stages of manufacture, the final product will have a bitter taste. On this account, a pure soft water supply is highly important to the industry.

Canneries and starch works as well as other food products manufacturers use large quantities of water and, of course, the first requirement is that it must be sanitary. In starch works, however, which manufacture glucose or corn syrup, the water must be unusually free from mineral salts because the dilute solutions have to be much concentrated in the course of boiling down the syrup. This causes any impurities which were originally present to appear in the finished product in much more concentrated form. Furthermore, if the glucose solutions are to be bleached by passage through charcoal filters, any mineral impurities which they contain will tend to deposit on the surface of the charcoal, thereby seriously impairing its efficiency. The same difficulty is met with in the manufacture of sugar. These industries, therefore, require a water supply of high purity.

Soap manufacturing uses large quantities of water in the soap-boiling operations and since calcium and magnesium salts precipitate an insoluble soap, it is obvious that this industry requires water low in hardness to produce a high-grade product. Likewise, the laundering industry, using large quantities of soap, will also require low-hardness water. The importance of soft water to all the industries using soap will be evident from the amount of soap destroyed by hardness. It is estimated that 18 to 20 pounds of soap are consumed per 1,000 gallons of water for every 100 parts per million of total hardness figured as calcium carbonate. This ratio will be even higher if the total hardness is low and it serves to show the high soap-consuming cost of hard water to these industries, to say nothing of the detrimental effect of the slimy precipitates on the value of the finished products.

Steam Generation

The chief industrial use of water aside from cooling purposes is in steam generation. Water for steam making is more or less suitable depending on the type and quantity of impurities which it carries. The customary method of interpreting the suitability of a water for boiler use is based on its tendency to cause foaming, corrosion, and scale or incrustation, and the cost of treatment to eliminate these tendencies is a fairly reliable inverse index of its value for this and many other industrial purposes.

Priming and foaming are probably the least understood of boiler phenomena. Priming may be defined as an ebullition so violent that water in the form of spray is carried from the boiler before its separation from the steam can take place. It is controlled by the relations of heating surface, evaporation surface, circulation, and working load, all of which are factors influencing the violence and rapidity of boiling, and by such features as dash plates, water space, and steam space, which affect the probability of spray reaching the steam exits. Priming, as thus defined, is a matter of boiler design and operation, unless the water is very high in soluble solids content.

Foaming is the formation of bubbles upon and above the surface of the water. The less easily these bubbles break, the higher will foam rise. It may become so excessive that the bubbles, or films of water enclosing steam, pass out with the steam. Naturally priming, or a tendency to prime, is an important factor in excessive foaming but, aside from this, the difficulty with which the steam pushes through the surface film of water and separates from it is a controlling agency.

With nearly pure water, foaming is very slight and never sufficient to cause the loss of water with steam in a well-designed boiler. Nearly all impurities dissolved or suspended in water increase the foaming tendency, though no two substances may do so to the same degree. As steam is used from the boiler, the impurities are concentrated and finally a stage is reached which will cause excessive foaming. The quantity of impurities and the effect of each influences the foaming tendency of a water.

Suspended solids have an influence on foaming tendency, but it is practically impossible to predetermine the quantity of suspended matter in a boiler at any time. Turbidity originally present in the water is largely precipitated, while additional suspended matter is derived from the precipitation by treatment agents of impurities in solution in the feed water. Although both these classes of substances are undoubtedly important, the effect of precipitated magnesium being especially noteworthy, their role in inducing foaming cannot be predicted from an analysis of boiler feed water.

The dissolved impurities in boiler water are usually regarded as the more important factors contributing to foaming. As the boiler water becomes more concentrated, the relative proportion of sodium and potassium salts present increases until all other dissolved substances become relatively insignificant in amount. It is, therefore, customary to attribute foaming primarily to sodium and potassium, since these substances are highly soluble and their relative importance in different waters is easily determined from analysis.

Corrosion is one of the most harmful effects of impurities in water and is due to free acid or to acid formed by the hydrolysis of salts in the water or, in rarer cases, to acid formed by the hydrolysis of fatty lubricating oils and to free oxygen. Of the harmful metal radicals, magnesium is the one most likely to occur in quantity in natural waters, and when this occurs, in conjunction with such negative radicals as chloride or sulfate, the water will probably be corrosive in proportion to the quantities present.

One of the first results in a boiler using untreated water is the precipitation of at least a part of the carbonate and bicarbonate radicals as calcium carbonate. Such precipitates can be acted upon by free acid, and to some extent by acids formed by hydrolysis, thus tending to neutralize them. The extent of this action is not well defined but it certainly offers some protection. More positive protection is usually secured by addition of alkali to the feed water to neutralize acidity and an antioxidant to remove free oxygen.

The formation of scale in boilers is a common effect of the use of impure untreated feed water. This phenomenon is the result of heating the water to a high temperature and concentrating it. The heat drives carbon dioxide out of the water, thus converting bicarbonate radicals to normal carbonate and causing the precipitation of insoluble carbonate salts, especially calcium carbonate. Concentration will gradually increase the amount of dissolved matter to saturation, after which additional evaporation will cause it to pass out of solution. Suspended matter and colloidal matter, especially silica, are also largely deposited within the boiler.

Almost all natural waters, if used in a boiler for a great length of time without cleaning, would produce scale or sludge. As boilers are usually operated, temperatures and concentrations are permitted which result in the precipitation of paretically all suspended and colloidal matter, all iron, aluminum, and magnesium, and all calcium to the full extent of its ability to combine with carbonate, and sulfate radicals. The iron, aluminum, and magnesium appear in the scale as oxides (magnesium carbonate may be present but is not likely to be found in quantity in scale from high-pressure boilers), while the calcium may be present as calcium carbonate or calcium sulfate (a hydrated calcium sulfate frequently occurs, but in the modern high-pressure boiler its quantity is sufficiently small to be neglected). These results may be caused by a series of reactions or by a single chemical change. The following reactions are presented for the changes which actually take place, and as equations which express the ultimate results of incrusting changes that occur in the boiler operated on untreated water.

$$2Fe^{-3} + 3H_2O = Fe_2O + O_3 + 6H$$
$$Mg^{-2} + H_2O = MgO + 2H$$
$$Ca^{-2} + CO_3 = CaCO_3$$

$$Ca^{-2} + 2HCO_3 = CaCO_3 + H_2O + CO_2$$
$$Ca^{-2} + SO_4 = CaSO_4$$

Scale may be of varying hardness, from a soft sludgelike deposit which can be flushed out, to a cement-like substance which must be chiseled out, depending on the character of the impurities. Colloidal matter, especially silica, contributes to dense scale formation, as does also calcium sulfate. Of these substances, the one which is likely to occur in serious quantities is calcium sulfate. This material, contrary to the general rule, is less soluble in hot than in cold water. Hence, as the temperature increases the calcium sulfate becomes less soluble and tends to come out in the form of scale aside from the effect due to the concentrating action of evaporation. In general, a water low in silica and sulfate will be low in hard scale-forming tendency and vice versa. The net value of a water for steam-making purposes will take into account the combined effects of foaming, corrosive, and scale-forming impurities.

WATER POLLUTANT CATEGORIES

Saline Water

Occasionally, salt brines from mines or oil wells are released into usually fresh water. While some organisms can tolerate a certain range of salt concentration, many disappear as fresh water becomes brackish. If the salt level remains constant, salt-tolerant species often appear and reproduce successfully. However, because of constantly fluctuating levels of most pollutants, organisms that might otherwise adjust to the new conditions rarely have the opportunity to become acclimated so that neither native nor newly introduced organisms can survive.

Another type of salt pollution has developed as the use of sodium chloride-calcium chloride mixtures on snowy roads in winter has increased to millions of tons per year. Although limited to states with winter snow problems, the quantities of salt use are staggering. The salt runoff changes the osmotic balance of soils bordering the highway and makes plants much more susceptible to drought. Sugar maples seem especially vulnerable to salt in the soil. Massive intrusions of salt into the water table from carelessly placed and unprotected storage piles have contaminated wells, and runoff from highways into storm sewers and natural watercourses is beginning to affect aquatic life in some areas. The chloride ion complexes strongly with mercury; the contamination of fresh water with deicing salt can release mercury from polluted bottom sediments.

Acid wastes not only lowers the pH of streams and contributes large quantities of iron and sulfate to the water but reduces the availability of oxygen in streams as well. In the low-water season of summer and early fall, pH may fall to 4.5, but with the high water of winter and spring, flushing of wastes from strip mines and mine shafts by tributary streams may lower the pH to 2.5. At pH levels below 4.0 all vertebrates, most invertebrates, and many microorganisms are eliminated. Most higher plants are absent, leaving only a few algae and bacteria to save the stream from total sterility. Acid mine drainage is a serious problem wherever coal has been mined. In the United States, the most severely affected areas are parts of Pennsylvania, Ohio, West Virginia, Kentucky, Illinois, Missouri, and Tennessee; in Europe, parts of the British midlands, France, Germany, and Poland. Most coal mining areas currently have strict regulations which have reduced pollution from such sources. Regulations, however, do not remediate those shafts or stripped areas already abandoned which continue to discharge acid wastes into the nearest drainage.

Although coal mining operations add the greatest amount of acid to streams, other industries add such wastes which affect pH. Continuous casting and hot rolling mills use sulfuric acid (pickling liquor) to clean oxides and grease from the processed metal. When the pickling solutions become diluted with iron salts and grease, they must be replaced with fresh solution. Since over 50 percent of all steel products are treated with pickling liquor, a considerable volume of spent liquor is disposed of every year. After pickling, the metal is rinsed to stop the pickling action. Because of their acidity these spent pickling and rinse solutions are a significant difficult wastewater problem.

Organic Pollutants

Common pollutants are usually not poisonous to stream life, nor do they necessarily affect pH. Most organic materials are attacked by bacteria and broken down into simpler compounds. Bacteria require oxygen to do this. The greater the supply of organic food, the larger the population of bacteria that can be supported, and the greater the demand on the oxygen in the water. This demand for oxygen by the bacteria is called biological oxygen demand (BOD). BOD is an index of pollution, especially related to the organic load of the water. Since all stream animals depend on the oxygen supply in water, the BOD is of particular importance in determining which forms of life a polluted stream is capable of supporting. Fish have the highest oxygen requirement; usually cold-water fish require more oxygen than warm-water fish. Invertebrates can tolerate lower oxygen concentrations and bacteria still lower.

Sometimes when the organic load is especially high and the water warm, oxygen is so depleted that the usual oxygen-requiring or aerobic decomposers are replaced by species that do not require oxygen (anaerobic), with quite different end products (see Table 1-3).

Methane (CH_4) is odorless. Amines have a distinct smell; hydrogen sulfide (H_2S), smells like rotten eggs, and some phosphate compounds have a wormy or rancid smell. Occasionally severe pollution from excess organic matter occurs in natural situations, such as heavy leaf fall temporarily overwhelming a small stream, creating a large BOD and possibly killing fish. But most organic wastes are related to man's misuse of the environment by the release of untreated human and animal wastes as well as those from industrial processing.

Table 1-3 End Products Under Differing Conditions

AEROBIC CONDITIONS	ANAEROBIC CONDITIONS
$C \rightarrow CO_2$	$C \rightarrow CH_4$
$N \rightarrow NH_3 + HNO_3$	$N \rightarrow NH_3$ + amines
$S \rightarrow H_2SO_4$	$S \rightarrow H_2S$
$P \rightarrow H_3PO_4$	$P \rightarrow PH_3$ and phosphorus compounds

Human Factors

No natural body of water is without life. Those with relatively small populations of plants and animals are called *oligotrophic*; their low level of fauna and flora results in very clear blue water. Lakes with high populations of aquatic life are termed *eutrophic*; because of the density of organisms, the water is turbid and green. Causes of the eutrophication are many and varied but generally can be traced to the abundance of three nutrients: phosphorus, nitrogen, and carbon.

The concentration of phosphorus in aqueous environments is normally low (about 0.02 parts per million) because most phosphates are insoluble in water. If the input of phosphorus in aquatic systems is increased much beyond this figure, it is either organically fixed by organisms and ultimately deposited in sediments, or precipitated directly. Excess phosphorus is incorporated into sediments eventually, but the pathways are of critical importance. When a nutrient normally in short supply is added to a system it may trigger a rapid growth of organisms in that system. Phosphorus is such a trigger factor in most aquatic systems. Microorganisms are often held in check by a lack of phosphorus, but when it is made available they use it immediately, and their rapid reproductive rates can suddenly become extreme.

Organic matter is a key to the eutrophication problem and carbon may often be more limiting than phosphorus. Adding large amounts of organic matter to aquatic systems stimulates a bloom of bacteria, which produces large quantities of free carbon dioxide. This in turn stimulates the rapid reproduction of algae that contributes to the overall eutrophication of the system.

Despite the quantity of phosphorus being added to the environment via detergents, the fertilization of farmlands has been a significant source. When highly soluble fertilizer is used, 10 percent to 25 percent is leached away into the surface runoff before the plants are able to use it. This represents a loss to the farmer as well as a burden to the aquatic ecosystems. Another agricultural source of phosphorus is animal waste. Livestock, increasingly grouped in feedlot, contributes significant phosphorus to the environment. Figure 1-4 illustrates the multiplicity of water uses, both agricultural and industrial as discussed in this chapter.

2 WATER QUALITY AND PROPERTIES

Strictly defined in the chemical sense, water is a compound, H_2O, and as all other pure substances, it has definite and constant composition. As any pure compound, water exhibits predictable chemical and physical characteristics. Properties of pure compounds are so dependable that they may be used for identification if an unknown sample is submitted to a laboratory.

One of the predictable physical properties of this widely distributed compound is its ability to dissolve numerous other materials. Compared with other known liquids, none of the others is capable of dissolving so wide a range of compounds of varying compositions. As a result, water seldom occurs in nature in a chemically pure state but typically contains impurities.

In addition to dissolved materials, water drawn from natural sources typically contains particles of insoluble, or undissolved, materials in suspension. The size and concentration of these suspended particles can vary considerably, depending upon the source, from sand grains sometimes present in rapid, turbulent surface streams to the submicroscopic dispersions or colloids. Included among the suspended particles, there may be living cells of different kinds of microorganisms. In the quality of water, concern is not really with the water itself, but with the other materials present. Impurities determine, to a very large degree, the suitability of a water source for its uses, the problems associated with utilizing it, and the kind and extent of treatment required.

GOALS OF WATER TREATMENT

In the broadest possible terms, the objectives of water treatment may be classified under three general requirements:

- to protect the health of the community
- to supply a product which is esthetically desirable
- to protect the property of the consumers

Protection of public health implies that the treated water must be free of microorganisms capable of causing human disease, and the concentrations of any chemical substances which are poisonous or otherwise harmful must be reduced to safe levels. Rarely do raw water supplies contain significant levels of toxic chemicals. But, more often than not, the microbiological quality of the water requires improvement or protection. This aspect of water treatment has progressed to the point that the physiological safety of public water supplies usually is taken for granted. In some parts of the world, it is considered necessary when visiting a strange city to carry a private supply of drinking water, or to inquire whether it is safe to drink from the local supply. In the United States, it is unquestionably a credit to the water treatment profession.

Esthetically water supply requires that the final product shall be as low as possible in odor, turbidity, and suspended solids, as cold as possible, and free from undesirable tastes and odors. Tastes and odors are highly subjective, and it may not be possible to produce a product which is equally pleasing to all consumers. However, strong, distinctive tastes and odors, as well as those which are disagreeable to a significant percentage of the population, are not desirable. The esthetic quality of water cannot be completely divorced from the question of public health, since objections to the taste, odor, color, and so on of a perfectly safe public supply may prompt consumers to use water from another source which is more attractive but which, due to a lack of protection, may be considerably more dangerous.

Property protection and its implications depend upon the purpose for which the water is intended. Requirements may, and occasionally do, vary among different consumers for the same supply. For domestic supplies, requirements are that the water will not be excessively corrosive to plumbing and other metal equipment, that it does not deposit troublesome quantities of scale, and that it will not stain porcelain plumbing fixtures. For industrial purposes, the requirements may be even more stringent. Generally speaking, public suppliers do not find it practical to meet the strict and sometimes varied requirements of industrial customers. Instead, they maintain a quality suitable for domestic consumption, and if necessary the industries provide their own treatment for specific requirements.

Due to the solvent action of water on all materials which it encounters, natural water contains salts of sodium, potassium, magnesium, calcium, and iron, together with organic matter extracted from decaying plant and animal debris. In addition, the water may carry suspended matter as fine clay, fragments of organic matter of plant and animal origin, and living microorganisms.

Hardness is caused by metal ions which can precipitate soap. Most hardness in water is due to the presence of calcium and magnesium ions, but soaps can be precipitated by other polyvalent ions such as iron, manganese, aluminum, zinc, and strontium. Hard waters are not desirable in domestic water supply because large amounts of soap are necessary to produce lather. Also scale formation at elevated temperatures makes hard waters undesirable for industrial boiler use (see section on boiler waters). *Hard* water or *soft* water may often be a subjective term. The degree of hardness has been classified in the following terms:

PPM Calcium Carbonate	Hardness Description
0-60	Soft
61-120	Moderately hard
121-180	Hard
Greater than 180	Very hard

Water containing appreciable amounts of calcium and magnesium compounds in solution is called hard water because of the chemical action of these compounds upon soap, forming an insoluble product. The use of hard water is objectionable in steam boilers. The calcium and magnesium salts deposit out in the boiler tubes, forming a stone-like layer on the inner walls of the tubes, called boiler scale. Scale is a poor conductor of heat and acts as an insulator. The result is an increase in the consumption of fuel. Frequently, when the fires are pushed, the boiler tube, separated from the water by the scale, becomes red hot and expands away from the scale. Under such conditions the scale may crack, and the water coming in contact with the red-hot metal is instantly converted into steam, producing an enormous pressure against the softened metal which bursts, and an explosion results.

Water Softening

Water which contains large quantities of calcium bicarbonate $Ca(HCO_3)_2$ is called temporary hard water. It may be softened by boiling. Calcium carbonate $CaCO_3$ precipitates and carbon dioxide escapes:

$$Ca(HCO_3)_2 \rightarrow CaCO_3 \downarrow + H_2O + CO_2 \uparrow$$

In many industrial plants water is softened by the addition of a sufficient amount of lime to precipitate the calcium as the carbonate:

$$Ca(HCO_3)_2 + Ca(OH)_2 \rightarrow 2\ CaCO_3 \downarrow + 2\ H_2O$$

Water containing calcium and magnesium sulfates cannot be softened by boiling. Such water is referred to as possessing permanent hardness and is softened by the addition of sodium carbonate:

$$CaSO_4 + Na_2CO_3 \rightarrow CaCO_3 \downarrow + Na_2SO_4$$
$$MgSO_4 + Na_2CO_3 \rightarrow MgCO_3 \downarrow + Na_2SO_4$$

In the resin-exchange process water is softened by simply filtering through a layer of coarse sand composed of an artificial sodium aluminum silicate. The sodium is replaceable by calcium, magnesium, ferrous, and other bivalent ions. The latter are removed from the water, being replaced by sodium ions, and form the insoluble permutite:

$$2 \text{ Na exchanger} + Ca^{++} \rightleftharpoons Ca \text{ (permutite)}_2 + 2 \text{ Na}^+$$

The equilibrium is reversible and when the removal efficiency fails, it may be regenerated by treating for a few hours with a concentrated sodium chloride solution. The process is inexpensive, since sodium chloride is consumed. Natural zeolites are used for this purpose.

Potable Water

Water which is fit to drink, containing nothing injurious to health, is called potable water. The mineral matter commonly found in water is not injurious; on the contrary, the presence of a certain amount may be advantageous, since it supplies a portion of the mineral constituents necessary for body growth and metabolism. Water should be free from suspended impurities and pathogenic or disease-producing bacteria. Typhoid and cholera epidemics have been traced to infected water supplies.

The purification of water for community use is an important practical problem for the engineer and chemist. The methods adopted vary with the local conditions. Suspended impurities are removed by filtration, employing large filtering beds of sand and gravel. Such treatment will not remove the bacteria. The addition of alum and lime, or ferrous sulfate and lime, produces a gelatinous, slimy coating on the grains of sand, which entangles and retains the bacteria along with very fine particles that would normally pass through the sand filter. This is called coagulation treatment.

When the water supply is free from suspended matter, the bacteria are killed by the addition of small amounts of chlorine, bleaching powder, ozone, and so on. Frequently, there are present in the water microscopic plants, known as algae, which can impart odor or color to the water. Such plants are destroyed by treating the water with cupric sulfate. Concentrations of the chemicals to be used must be such that their effect upon the higher forms of life is negligible, while the lower forms are destroyed. Sometimes the water is impounded and filtered through covered filters to prevent the growth of algae which require sunlight. Ultraviolet light, produced by passing an electric arc through mercury vapor in a quartz tube, may be employed for killing the colon bacilli organisms associated with those which produce typhoid fever by allowing the rays to penetrate the water.

Household water which is suspect may be rendered safe by boiling. Most organisms are killed by boiling for 10 or 15 minutes. Such water must be carefully preserved to avoid contamination, since bacteria are present in the air and the water may contain nitrogenous organic matter upon which the bacteria live and propagate. The bacteria may also be removed by the use of a household filter, provided the pores of the filtering medium are sufficiently small. A Pasteur filter made of special unglazed porcelain has exceedingly small pores. When this is attached to the faucet, the water is forced through under its own pressure and the suspended matter and bacteria are held back, forming a slimy coating on the porcelain. The filter must be cleaned daily with a brush to remove this deposit, otherwise the bacteria multiply and eventually the filter pollutes the water instead of purifying it.

Regarding the use of ice in drinking water, natural ice, particularly that which has been stored for four to eight months, is quite safe. During the process of freezing, suspended and dissolved matter is not removed from water with the ice, except a small amount which is mechanically enclosed within the mass of crystals. Over 90 percent of the pathogenic bacteria die during long storage. Ice made from polluted water is not safe because all of the suspended matter is sometimes frozen into the center of the cake. Ice made from potable or from distilled water is perfectly safe for all purposes.

Chemical and Bacteriological Examination

Following are three factors considered in determining the potability of a water and accepting it as a supply for a municipality.

- Regular bacteriological examination should be made to determine the presence of pathogenic bacteria. It is important to know the number and the types of bacteria present in order to prescribe the proper treatment.
- Frequent chemical examination of the water should be made to determine the presence of color, odor, and nitrogenous organic matter. The color is usually caused by organic matter of a harmless nature, but a water with excess color is not esthetically attractive. The color is frequently removed by coagulation treatment. The odor may be due to the presence of inorganic gases (hydrogen sulfide, sulfur dioxide, and so on). Such water will not cause disease, but may have some physiological action on persons not accustomed to their use. However, odors may be due to sewage contamination and in such cases, the use of that water must be prohibited until the proper treatment is effected. A chemical examination may show the presence of nitrogenous organic matter, for example, albuminoids, upon which bacteria feed; or free ammonia, nitrites, and nitrates, the products resulting from bacterial life, which mean that bacteria may be present. If the analysis does not prove that the bacteria present are disease producing, such water is to be suspected and should be avoided nevertheless.
- A regular systematic survey should be made of the sources from which water is obtained. By such investigation a temporary or a permanent source

of contamination can be located and the proper steps taken to avoid the evil consequences arising from the use of the water.

To maintain the quality of a water treatment plant various chemical and physical tests must be performed (See Chapter 4). This assures meeting the standards which are required and desired. If for any reason the quality is temporarily unsatisfactory, test results advise of the problem, and permit corrective action. By keeping permanent records of the results, the demonstrated quality of the water to regulatory authorities or to other interested individuals or agencies can be made.

Bacteriological quality is most often based on measurements of the numbers of coliform bacteria. Although this group of organisms is not known to cause human disease directly, its presence and survival is considered to indicate the potential presence of disease organisms (pathogens). Consequently, the number of coliforms present is strictly regulated. The enumeration of coliforms is supplemented by the *total plate count*, which is an approximate measurement of the total microbial population of the water, or by determining the numbers of one particular species of the coliform group, Escherichia coli.

In the majority of water treatment plants, control of the bacteriological quality of the water is effected by means of chlorination. Determination of residual chlorine in its various forms becomes an important analysis, even though it may not be rigorously correct to consider it a direct means of monitoring the water quality. Related to the measurement of residual chlorination is the determination of chlorine demand, which is defined as the difference between the concentration of chlorine added and the concentration remaining after a specified period of time. Measurement of the chlorine demand of the raw water is often essential to successful control of the bacteriological quality of the finished product, particularly if the chlorine demand of the source tends to be variable.

Tests for chemical substances known to be toxic are not ordinarily conducted routinely unless there is reason to suspect their presence. If the previous history of the water supply or other circumstances indicate a possibility of this kind, the analytical program should include measurement of the concentration of the substances, probably both before and after treatment. Otherwise tests of this type might be included among those which are performed only periodically.

Tests related to the esthetic quality of the water, determinations of color, turbidity, suspended solids, and temperature are important. Measurement of taste and odor is almost as subjective in the laboratory as in the consumer's home or place of business. The determinations are not routinely performed but rely upon complaints to advise of the occurrence of a problem. Where strong or disagreeable tastes and odors are a frequent problem, such tests may be a regular part of the quality control program. Substances such as sulfides and phenols which are known to impact taste and odor may be measured. Also, the determination of iron and manganese may be included because excessive quantities of either may affect both taste and color. Measurement of dissolved oxygen is sometimes included since the majority of people seem to prefer the flavor of water in which the oxygen content is near saturation.

For domestic supply, analyses related to the protection of property include the tendency of the water to corrode metals or to deposit scale. Important tests are those for pH, acidity, alkalinity, total hardness, and calcium. Sometimes a determination of conductivity and total solids may be included, and under certain circumstances a measurement of the concentration of sulfates is important.

Drinking Water Standards

The U.S. Department of Health, Education and Welfare, through the U.S. Public Health Service, publishes standards for the quality of drinking water. Although the federal Public Health Regulations govern interstate carriers and certain other specified installations, their standards are widely used as guide by other regulatory agencies. Many of the latter have incorporated the Public Health Service standards wholly or in part into their own rules.

The standard of bacteriological quality is based upon the number of coliform bacteria present. Detailed sampling and testing procedures are specified, and a complete and fairly elaborate description of the method of evaluation sets forth precisely what results are required of an acceptable supply. In effect, the number of coliform bacteria is limited to not more than one organism per 100 ml of water on the average, with not more than 5 percent of the samples tested showing numbers greater than this limit.

With regard to physical properties, the turbidity should be less than five units, the color less than 15 units, and the threshold odor number less than three. If the turbidity standard is satisfied, the suspended solids will not be detectable.

Radioactivity is limited, but the acceptability of a given supply is dependent to some extent upon exposure from other sources. A water supply is unconditionally acceptable in this respect if the content of radium 226 is less than three micro-micro-curies per liter, the content of strontium 90 is less than 10 micro-micro-curies per liter, and the gross beta-ray activity is less than one microcurie per liter. If the radioactivity of the water supply exceeds the values stated, then its acceptability is judged on the basis of consideration of other sources of radioactivity in the environment.

WATER COMPOSITION

All water used to supply human requirements has at some time fallen to the surface of the earth as rain or some other form of precipitation. At this stage, the quantity of foreign material it contains is likely to be minimum. However, even rain water is not chemically pure H_2O. Not only does it dissolve the gases of the atmosphere as it falls, but it also collects dust and other solid materials suspended in the air. Since the atmospheric solids depend upon both the composition of the soil below and the materials released into the air from combustion, industrial processes, and other sources, analyses of rain or other forms of precipitation reveal variations. In general, however, rainwater may be expected to be soft, to be low in total solids and alkalinity, to have a pH value somewhat below neutrality, and to be corrosive to many metals. A typical analysis, subject to variations mentioned, might be as follows:

Hardness	19	mg/l as $CaCO_3$
Calcium	16	mg/l as $CaCO_3$
Magnesium	3	mg/l as $CaCO_3$
Sodium	6	mg/l as Na
Ammonium	0.8	mg/l as N
Bicarbonate	12	mg/l as $CaCO_3$
Acidity	4	mg/l as $CaCO_3$
Chloride	9	mg/l as Cl
Sulfate	10	mg/l as SO_4
Nitrate	0.1	mg/l as N
pH	6.8	

After the water reaches the surface of the ground, it passes over soil and rock into lakes, streams, and reservoirs, or percolates through the soil and rock into the groundwater. In the process, a great variety of materials may be dissolved or taken into suspension. It may be expected that the composition of both the surface waters and groundwater of a given area reflects the geology of the region, that is, the composition of the underlying rock formations and of the soils they are derived from. The presence of readily soluble formations near the surface such as gypsum, rock salt, or the various forms of limestone produce relatively marked effects upon the waters of the area. In the presence of less soluble formations, such as sandstone or granite, the composition of the water tends to remain more like that of rain. Local variations are often considerable and can occasionally be extreme, both in the concentration of any one constituent and in the proportions of the various materials present.

Surface water in an area in which limestone is an important constituent of the geologic formations might have a composition as follows:

Hardness	120	mg/l as $CaCO_3$
Calcium	80	mg/l as $CaCO_3$
Magnesium	40	mg/l as $CaCO_3$
Sodium and Potassium	19	mg/l as Na
Bicarbonate	106	mg/l as $CaCO_3$
Chloride	23	mg/l as Cl
Sulfate	38	mg/l as SO_4
Nitrate	0.4	mg/l as N
Iron	0.3	mg/l as Fe
Silica	18	mg/l as SiO_2
Carbon Dioxide	4	mg/l as $CaCO_3$
pH	7.8	

In such a source, the groundwater may often contain more hardness and bicarbonate than the surface waters. This is due to the longer period of contact with soil and rock, and to the fact that carbon dioxide, contributed by the decomposition of organic matter in the soil, greatly increases the solubility of some of the constituents. The following might be considered typical of well or spring water in a limestone area:

Hardness	201	mg/l as $CaCO_3$
Calcium	142	mg/l as $CaCO_3$
Magnesium	59	mg/l as $CaCO_3$
Sodium and Potassium	20	mg/l as Na
Bicarbonate	143	mg/l as $CaCO_3$
Chloride	23	mg/l as Cl
Sulfate	59	mg/l as SO_4
Nitrate	0.06	mg/l as N
Iron	0.18	mg/l as Fe
Silica	12	mg/l as SiO_2
Carbon Dioxide	14	mg/l as $CaCO_3$
pH	7.4	

In areas in which the underlying formations are insoluble, that is, where they consist of sand, sandstone, clay, shale, or igneous rocks, the waters tend to be softer and more acid. In general, their content of most dissolved materials is lower. Acidity, however, may be higher than in hard-water areas, since carbon dioxide picked up from the soil is not neutralized. Excepting in some areas of igneous rock, iron also tends to be higher in soft waters, since many of the iron compounds of soils and rocks are dissolved by the acidity of the waters. In many soft-water areas, the differences between groundwaters and surface waters are not as pronounced as in hard-water regions, although many exceptions to this generality could be cited.

A typical analysis of surface water in a region of generally insoluble soils and rocks is:

Hardness	46	mg/l as $CaCO_3$
Calcium	30	mg/l as $CaCO_3$
Magnesium	16	mg/l as $CaCO_3$
Sodium and Potassium	9	mg/l as Na
Bicarbonate	42	mg/l as $CaCO_3$
Chloride	5	mg/l as Cl
Sulfate	12	mg/l as SO_4
Nitrate	1.5	mg/l as N
Iron	1.1	mg/l as Fe
Silica	30	mg/l as SiO_2

Groundwater from a similar region might give analyses similar to:

Hardness	61	mg/l as $CaCO_3$
Calcium	29	mg/l as $CaCO_3$

Magnesium	32	mg/l as $CaCO_3$
Sodium	26	mg/l as Na
Bicarbonate	60	mg/l as $CaCO_3$
Chloride	7	mg/l as Cl
Sulfate	17	mg/l as SO_4
Carbon Dioxide	59	mg/l as $CaCO_3$
pH	6.6	
Iron	1.8	mg/l as Fe

Each of the constituents listed in the preceding analyses may vary over a wide range from location to location. For example, waters are known with hardness values of less than 10 mg/l, and others have concentrations over 1,000 mg/l. No water sample will correspond exactly to any one of the analyses given in the preceding examples.

Since waters from various sources may vary so widely in composition, which source should be considered most desirable? For example, if a choice exists among several available sources, the final decision may rest upon the judgment of relative quality between them. When the composition is modified by treatment, the objective is to approach, if not always to attain, the ideal.

Properties of good water in qualitative terms are as follows:

- Absence of harmful concentrations of toxic chemical substances
- Absence of the microorganisms and viruses that may cause disease
- Lowest possible levels of color, turbidity, suspended solids, odor, and taste
- Lowest possible temperature
- Minimum corrosion to metals
- Least possible tendency to deposit scale
- Lowest possible content of staining materials, such as iron, manganese, and copper

Extremely soft waters tend to be excessively corrosive to metals and may be found unpalatable. Moreover, they may be less effective in removing soap by rinsing than waters containing a little hardness. Although there has been no formal recognition of a set of analytical values characterizing the ideal water, the following would probably be considered generally acceptable:

Alkyl Benzene Sulfonate	less than 0.1 mg/l, preferably 0
Arsenic	less than 0.01 mg/l, preferably 0
Barium	less than 1 mg/l, preferably 0
Bicarbonate	150mg/l as $CaCO_3$
Cadmium	less than 0.01 mg/l, preferably 0
Calcium	70 mg/l as $CaCO_3$
Carbon Chloroform Extract	less than 0.2 mg/l, preferably 0
Carbon Dioxide	6mg/l as $CaCO_3$
Chloride	less than 250 mg/l, preferably 0

Chromium, Hexavalent	less than 0.05 mg/l, preferably 0
Coliform Bacteria	less than 1 per 100 ml
Color	less than 15 units, preferably 0
Copper	less than 1 mg/l, preferably 0
Cyanide	less than 0.01 mg/l, preferably 0
Fluoride	approximately 0.9 mg/l (somewhat dependent upon climate)
Hardness	70 mg/l as $CaCO_3$
Iron	less than 0.1 mg/l, preferably 0
Lead	less than 0.05 mg/l, preferably 0
Magnesium	preferably 0
Manganese	less than 0.02 mg/l, preferably 0
Nitrate	less than 10 mg/l, preferably 0
pH	7.8
Phenols	less than 0.001 mg/l, preferably 0
Selenium	less than 0.01 mg/l, preferably 0
Silver	less than 0.05 mg/l, preferably 0
Sodium and Potassium	37 mg/l as Na
Sulfate	less than 250 mg/l, preferably 0
Suspended Solids	not detectable
Temperature	33 to 40° F
Threshold Odor Number	less than 3, preferably 0
Total Dissolved Solids	less than 500 mg/l
Turbidity	less than 5 units, preferably 0
Zinc	less than 5 mg/l, preferably 0

Relationships among calcium, bicarbonate, carbon dioxide, and pH should be such as to minimize scaling and corrosion. In some cases, these concentrations may dictate the most desirable concentrations of sulfate, chloride, magnesium, sodium, and potassium.

Solubilities of some inorganic compounds in water at 18°C (64.4°F) are shown in Table 2-1.

Physical Properties

At ordinary temperatures, water in relatively thin layers is a clear, transparent, colorless liquid. Deep layers of water, such as lakes, have a bluish-green appearance. The freezing point and the boiling point of water are taken as 0.000°C. and 100.000°C, respectively, at 760 mm pressure. At 4°C water possesses its maximum density; 1 ml of water at 4°C weighs 1 gram. When heated above or cooled below this temperature, water expands and the density decreases, as follows:

Table 2-1 Solubilities of Inorganic Compounds in Water at 18° C (64.4°F)

A = number of grams of the anhydrous compound dissolved in 100 ml of water.
B = molar solubility, i.e., the number of moles contained in 1 liter of the saturated solution.

SOLUTE	A	B	SOLUTE	A	B
Barium carbonate	0.0022	$0.0_3 11$	Potassium hydroxide	110.	14.6
Barium chloride	35.2	1.7	Potassium iodide	142.	6.2
Barium chromate	$0.0_3 35$	0.014	Potassium nitrate	29.6	2.4
Barium hydroxide	3.7	0.216	Potassium sulfate	11.1	0.6
Barium iodide	201.36	3.8	Silver bromide	$0.0_4 1$	$0.0_6 6$
Barium sulfate	$0.0_3 23$	$0.0_4 1$	Silver chlorate	12.25	0.6
Calcium bromide	143.	5.2	Silver chloride	$0.0_3 15$	$0.0_5 9$
Calcium carbonate	$0.0_2 13$	$0.0_3 1$	Silver chromate	0.0025	$0.0_4 75$
Calcium chloride	72.	5.26	Silver fluoride	195.4	13.5
Calcium chromate	2.3	0.147	Silver iodide	$0.0_6 35$	0.071
Calcium fluoride	$0.0_2 16$	0.032	Silver nitrate	211.6	8.3
Calcium hydroxide	0.17	0.023	Silver sulfate	0.58	0.02
Calcium iodide	202.8	4.86	Sodium bromide	89.1	6.9
Calcium nitrate	125.8	5.37	Sodium carbonate	19.3	1.8
Calcium sulfate	0.20	0.014	Sodium chlorate	97.16	6.4
Lead chloride	1.49	0.05	Sodium chloride	35.95	5.4
Lead iodide	0.08	0.002	Sodium fluoride	4.49	1.07
Lead sulfate	$0.0_2 41$	0.031	Sodium hydroxide	107.5	20.0
Lithium carbonate	1.3	0.17	Sodium iodide	177.9	8.1
Lithium hydroxide	12.04	5.	Sodium nitrate	86.1	7.5

Table 2-1 (continued)

SOLUTE	A	B	SOLUTE	A	B
Magnesium carbonate	0.1	0.011	Sodium sulfate	16.83	1.15
Magnesium chloride	56.	5.1	Strontium carbonate	0.0011	0.047
Magnesiumhydroxide	0.001	0.02	Strontium chloride	52.7	3.1
Magnesium sulfate	35.2	2.8	Strontium hydroxide	0.77	0.063
Potassium bromide	65.8	4.6	Strontium sulfate	0.011	0.0_36
Potassium carbonate	111.	6.3	Zinc hydroxide	0.0_34	0.0_44
Potassium chlorate	6.6	0.52	Zinc nitrate	115.	4.6
Potassium chloride	33.9	4.0	Zinc sulfate	55.6	3.4

NOTE: The numbers for small solubilities have been abbreviated, thus, $0.0_7 = 0.00007$.

TEMPERATURE	DENSITY	TEMPERATURE	DENSITY
0°C	0.9998	40°C	0.9923
4°C	1.0000	60°C	0.9833
10°C	0.9997	80°C	0.9719
20°C	0.9982	100°C	0.9586

Water has a higher heat capacity than any other substance except hydrogen. Specific heat of a substance is defined as the amount of heat necessary to raise the temperature of 1 gram of that substance 1 degree Centigrade; the specific heat of water is 1 calorie, while the specific heat of all other substances, except hydrogen, is less than 1 calorie. This accounts for the fact that large bodies of water, such as lakes and the ocean, change in temperature more slowly than the rocks and soil making up the land, and tend to regulate the temperature of the air by either absorbing large amounts of heat as the water warms up during the summer months, or liberating heat as it cools during the colder seasons. This phenomenon is noticeable on islands in the ocean where the climate is less variable from season to season than the climate on a continent.

When heat is removed from water at 0°C, it changes into ice. To change 1 g of water to ice at 0°C requires the removal of 79 calories. To change 1 g of ice to water at the same temperature, 79 calories are absorbed. This amount of heat is called the heat of fusion of ice. A mixture of ice and water at 0°C will remain unchanged in proportions unless heat is added or removed. When heat is added slowly, some of the ice will melt and absorb the heat. The temperature of the mixture, however, will not change until all of the ice has melted. Similarly, as heat is removed, water will freeze to ice and give up heat, but the temperature of the mixture will not decrease until all of the liquid has changed to ice. The temperature at which the liquid is in equilibrium with the solid is called the melting or freezing point. An equilibrium mixture of ice and water is used as a fixed temperature for calibrating instruments for measuring temperatures; for example, the zero on the Centigrade scale.

When water changes into ice at 0°C, it undergoes considerable expansion and change in density so that ice floats in water. This is a rather unusual phenomenon, since nearly all other substances contract on changing from the liquid to the solid state. If this were not so for water, our lakes and rivers would freeze at the bottom and would probably remain frozen even during the summer months, thus killing all marine life. The density of ice at 0°C is 0.917.

Water, although not a typical liquid in all respects, can be used to illustrate properties of the liquid state. This state of matter differs from the gaseous state in that a definite amount of a liquid substance will occupy a volume that changes only slightly with a change of temperature or pressure. If the container is larger than the volume occupied by the liquid, there will be a free surface on the liquid. Like gases, liquids are fluid and will flow. Compared with gases, however, liquids are relatively incompressible. Thus, a pressure of 150,000 lbs per square inch on water will compress it only to about four fifths of its original volume. Because of this relative incompressibility, the low degree of expansion in volume with rise in temperature, and the enormous increase in volume when a liquid substance changes into a gas (steam occupies 1,700 times the volume of the liquid water from which it was formed), in the liquid state the molecules are quite close together. Liquids flow and take the shape of the containing vessel.

That molecules exist in a liquid and they have kinetic energy which varies with the absolute temperature was proved by the investigations on Brownian movement. Kinetic energy of the molecules accounts for the fact that some liquids diffuse into others, just as gases diffuse into each other. The attraction of molecules in a liquid for others is exerted in all directions and results in a balance of forces except when the molecule is at the surface of a liquid. A molecule in the surface layer is attracted by molecules within the liquid being pulled into the liquid and it is also attracted by other surface molecules. Unbalanced attractive force accounts for the phenomenon of surface tension. The surface of a liquid appears to be under a tension; it acts as though it were a stretched membrane. In water this surface membrane is tough enough to support a steel needle, and many insects can walk on such water surfaces. It is the tendency to contract the surface that causes a drop of liquid to take a spherical shape that forms with the minimum surface.

Evaporation

Water in an open container dries up. The liquid passes off as vapor and the phenomenon is called evaporation. There are three factors which affect the rate at which a given quantity of water will evaporate into the air:

- the area of the exposed surface of the gas liquid
- the temperature of the liquid and the air above it
- the movement of the air above the surface of the liquid

Liquids and gases are composed of molecules in motion. In liquids the movement of the molecules is impeded by the force of cohesion between the molecules and by molecular collisions. The molecules move about in all directions, colliding with and rebounding from the walls of the containing vessel, and from the surface of the liquid. Of those which come to the surface, some possess sufficient kinetic energy to overcome the cohesive force and break away from the liquid, becoming gaseous molecules in the space above the liquid. This breaking away from the liquid is evaporation, and takes place only at the surface of the liquid. As we increase the surface area, the number of molecules leaving the liquid in unit time increases.

The average kinetic energy of molecules increases with a rise in temperature. Hence, as the temperature of a liquid rises, the velocity of the molecules increases and the number leaving the liquid in unit time also increases. The molecules which leave the liquid and become gaseous move about with greater freedom. Some return to the liquid and condense. When a condition of equilibrium is reached, the number of molecules leaving the liquid equals the number returning in any unit of time. When this condition is reached, there is no cessation of activity, but both processes of evaporation and condensation go on with the same speed. The air above the liquid is saturated with the vapor of the liquid. In order to promote evaporation of the liquid, this saturated layer of air must be removed and replaced by air which is unsaturated so that the number of molecules returning to the liquid will be less than the number leaving the liquid. This is accomplished by constant change of the air above the liquid. The lower the concentration of water molecules in the air above the liquid, the fewer will be the number of returning molecules, and hence the faster we remove the partially saturated air the more rapid will be the evaporation.

Vapor Pressure

When a liquid evaporates, its vapor exerts a pressure of its gaseous molecules. A gas at a temperature below its critical temperature is often called a vapor.

A convenient way of measuring the vapor pressure of a liquid involves the use of a barometer tube. A small amount of the liquid is introduced below the tube so that it rises through the mercury column into the space above the mercury in the tube. The liquid evaporates very quickly into the vacuum and the pressure exerted by the vapor causes the mercury level to drop. An excess of the liquid must be used in order to obtain equilibrium. The drop in the level of the mercury column is the vapor pressure of the liquid at the temperature of the experiment. By surrounding the barometer tubes with jackets for maintaining constant temperatures, the vapor pressure may be determined at any desired temperature. Figure 2-1 shows the results obtained for various liquids. Tube 1 represents the height of the mercury column in each of the tubes at the start of the experiment. Tubes 2, 3, and 4 show the results obtained for water at 20°C, 50°C, and 100°C, respectively. Tubes 5, 6, 7, and 8 show the results for several liquids. The vapor pressure is indicated in millimeters of mercury. The vapor pressures for water (or aqueous tensions) at temperatures from 0°C to 100°C are given in Table 2-2.

Freezing Mixtures

In the following table, Column A is the proportion of the substance named in the first column to be added to the proportion of the substance given in Column B. The table gives the temperature of the separate components and the temperature attained by the mixture.

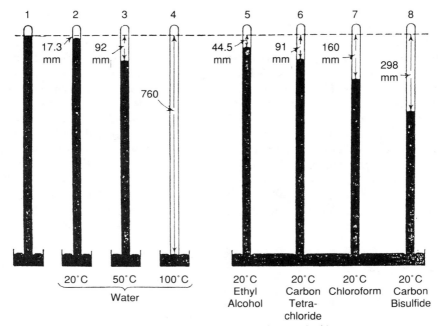

Figure 2-1 Vapor pressure determination

Substance	A	B	Initial	Temperature of Mix(°C)
NH_4Cl	30	H_2O 100	13.3	− 5.1
KI	140	H_2O 100	10.8	− 11.7
NH_4Cl	25	Snow 100	−1	− 15.4
NaCl	33	Snow 100	−1	− 21.3
$H_2SO_4 + H_2O$ (66.1% H_2SO_4)	1	Snow 4.32	−1	− 25.0
$CaCl2 + 6 H_2O$	1	Snow 0.61	0	− 39.9
	1	Snow 0.70	0	− 54.9
	1	Snow 0.81	0	− 40.3
Alcohol	---	CO_2 solid	---	− 72.0
Chloroform	---	CO_2 solid	---	− 77.0
Ether	---	CO_2 solid	---	− 77.0
Liquid SO_2	---	CO_2 solid	---	− 82.0

Water Quality and Properties 57

Table 2-2 Vapor Pressure of Water

Temperature °F	°C	Pressure mm	Temperature °F	°C	Pressure mm
32.0	0	4.6	78.8	26	25.1
41.0	5	6.5	80.6	27	26.5
46.4	8	8.0	82.4	28	28.1
48.2	9	8.6	84.2	29	29.8
50.0	10	9.2	86.0	30	31.5
51.8	11	9.8	87.8	31	33.4
53.6	12	10.5	89.6	32	35.4
55.4	13	11.2	91.4	33	37.4
57.2	14	11.9	93.2	34	39.6
59.0	15	12.7	95.0	35	41.8
60.8	16	13.6	104.0	40	55.0
62.6	17	14.5	122.0	50	92.2
64.4	18	15.4	140.0	60	149.2
66.2	19	16.3	158.0	70	233.8
68.0	20	17.4	176.0	80	355.5
69.8	21	18.5	194.0	90	526.0
71.6	22	19.7	212.0	100	760.0
73.4	23	20.9	302.0	150	3,581.0
75.2	24	22.2	392.0	200	11,588.0
77.0	25	23.6	446.0	230	20,925.0

Dissolved Gases

Dissolved gases, carbon dioxide (CO_2), hydrogen sulfide (H_2S), and oxygen (O_2) are not often determined because to be of value, the tests must be made at the source with special test equipment. Gas content in stream waters is necessarily small, but in groundwaters substantial quantities of gases may be held in solution and their importance may be significant.

Dissolved Solids

Dissolved solids are usually determined in number of parts per million multiplied by the reaction coefficient. This will give milli equivalent weights. Alkali radicals are frequently estimated and reported as sodium, the separation of the other radicals of the group being omitted. Since sodium may be 75 percent or more of the group, and several members have similar significance, this practice leads to no serious error in most industrial applications. Calcium and magnesium are the only members of the alkaline-earth group ordinarily determined; others, where present, are found in relatively small or insignificant proportions. Manganese, iron, and aluminum are often determined together as oxides and as reported as such. Acidity is reported in terms of hydrochloric acid, sulfuric acid, or calcium carbonate. Chloride and sulfate radicals are usually determined, but other strong-acid radicals, which are rarely in sufficient proportion to be of great industrial importance, may generally be omitted in untreated water. The weak-acid radicals may be determined together by the *alkalinity* titration, although separate determination of free carbon dioxide (CO_2), half-bound carbon dioxide (HCO_3), and fully bound carbon dioxide (CO_3) are often made. When the other weak-acid radicals are absent or nearly so, as is usually the case, such analyses are satisfactory. A determination of total solids is also frequently made and serves as a material balance on the sum of solid constituents computed from the individual analyses. There are a large number of instruments for analysis, giving direct readings of qualities of water such as total hardness, temporary hardness, and incrustants.

Water Supply Examination

The value of a water for domestic or industrial use depends upon its purity. Examination, therefore, consists of a determination of the nature and quantity of the impurities present. The type of examination will depend on whether the water is to be used for sanitary or industrial purposes.

Special sampling methods have been developed to insure representative samples for analytical purposes. For microscopic study, the organisms are gathered with as little disturbance as possible and it is recognized that these organisms are usually present in much higher concentration near the surface of the water. For bacteriological examination the usual precautions to prevent contamination with unsterile apparatus are taken. Further development of bacteria in the sample after it is collected is avoided by refrigeration and by reducing to a minimum the time interval between collection and analysis. For chemical examination where dissolved solids are examined, it is often necessary to collect the samples at a definite depth, and special devices have been designed to be lowered to any predetermined distance and then opened for obtaining the sample. Where dissolved gases must be determined, devices are used to insure a thorough flushing out of the sample bottle before the accepted sample is taken (see Chapter 4).

Sanitary Analysis

Sanitary analysis of water ordinarily consists of physical examination for temperature, turbidity, color, and odor; microscopical examination for number and kinds of microscopic organisms; bacteriological examination for total number and number of certain kinds of bacteria; and chemical examination, usually including total residue on evaporation, loss on ignition, fixed solids, alkalinity, hardness, chlorine, iron, nitrogen as albuminoid ammonia, nitrogen as free ammonia, nitrogen as nitrites, nitrogen as nitrates, and sometimes dissolved oxygen. Sanitary examination in general is of less industrial value than mineral analysis, although for water used in the food industries and for the control of purification plants, sanitary analysis is of prime importance.

Mineral Analysis

Mineral analysis of waters ordinarily used for industrial purposes includes four classes of water impurities: suspended solids, colloidal matter, dissolved gases, and dissolved solids.

Suspended Solids

Suspended solids include all organic or inorganic matter that can be removed by filtering. Such material is complex and can consist of many chemical compounds that may or may not be determined in detail. Two determinations often made: turbidity, the figures for which indicate the concentration of a known standard suspended matter that will obscure just as much light as the water under consideration; and suspended matter, which is the proportion by weight of the suspended solids in the water. The ratio depth of suspended matter to turbidity is called the coefficient of fineness. The greater the value of this coefficient, the greater will be the average weight of the suspended particles, and hence the greater will be the ease with which they can be removed.

Colloidal Matter

Colloidal matter includes, for the most part, silica (SiO_2), alumina (Al_2O_3), and iron oxide (Fe_2O_3), though in some waters, especially polluted waters, a considerable amount of organic matter may be present in the colloidal state. In a mineral analysis, colloidal matter is not distinguished from that in true solution. Colloidal matter, which is not in true solution, is so finely distributed that it does not usually contribute to turbidity. However, in a water which contains no other suspended matter the presence of colloidal material may sometimes be detected by reason of a faint opalescence observable in brilliant light. Blue color haze observable in the almost crystal clear depths of the springs is attributed to material dissolved by the water under great pressure and released as colloidal matter as the water rises to the surface outlet.

PHYSICAL PROPERTIES

Density

Density is the relationship of the mass (M) of any given material to the volume (V) occupied by that material. This can be expressed by the formula $D = M/V$. According to this equation, density can be changed by altering the weight (mass) or volume of a substance. For example, if a liquid has a mass of and occupies a volume of 100 liters, its density would be 1 ($D = M/V$ or $D = 100/100 = 1$). If, on the other hand, the volume remains constant at 100 liters and the mass is decreased to 50 g, the density would decrease to 0.5 ($D = 50/100 = 0.5$). Thus, the density of a particular substance will increase if the mass is increased and the volume kept constant or if the volume is decreased while the mass remains constant (or increases). Conversely, the density will decrease if the mass is decreased and the volume held constant or increased, or if the volume is increased and the mass is held constant or decreased.

Temperature

The significance of temperature variations in water are:

- Physical form, for example, at 0°C (32°F) pure water becomes solid (ice). At 100°C (212°F) water boils and becomes vapor or gas (steam).
- Solubility of dissolved substances is generally directly proportional to the temperature in the liquid state.
- Vapor pressure of water increases from 0 mm at 0°C (32°F) to 760 mm Hg at 100°C (212°F). Table 2-2 shows the relationship of the vapor pressure of water to its temperature.

Water will tend to increase in density as it is cooled until it reaches its temperature of maximum density, maximum hydrogen bond formation, and minimum volume at 4°C.

As water is cooled below 4°C, it goes through its pseudocrystalline states, during which the molecules rearrange, spread out, and increase in volume prior to entering the open crystalline ice structure. Therefore, ice has a lower density than liquid water at any temperature. As water above 4°C is cooled, the density increase that is noted is due to reduced thermal agitation. This causes the water to move more slowly, thereby increasing the possibility of hydrogen-bond formation and the decrease in volume. As water is cooled below 4°C, the reduction in volume does not compensate for the increased numbers of water molecules entering their pseudo and truly crystalline forms. It is this increased number of these molecules that is responsible for the increased water volume occupied by water molecules at temperatures below 4°C.

Universal Solvent

In addition to its influence on density, polarity of the water molecule is the reason water is termed the universal solvent. If an ionic compound such as sodium chloride (NaCl) is placed in contact with water it readily dissolves and electrons are transferred from the Na atom to the Cl atom in the formation of the molecule. The sodium has, therefore, assumed a positive charge since it has lost an electron to the Cl atom, which has developed a negative charge in the process. Thus, the molecule may be drawn as $Na+ \rightarrow Cl-$. When a polar liquid such as water comes in contact with ionic molecules or polar compounds, the water will orient itself around these molecules so that the positive ends of the water align with the negative regions of the molecule, and the water's negative regions align with the molecule's positive portions.

In this intermolecular attraction between the water, the solvent, and NaCl, the solute tends to break the chemical bond between the sodium and the chlorine, and the molecule separates into its ionic components. This separation is dissociation. When the bond breaks, the electron that sodium lost in the initial NaCl formation remains with the chlorine. Therefore, sodium goes into solution as a positively charged particle (ion) and chlorine as the negatively charged particle or the chloride ion.

In determining a given solute's ability to dissolve in a given solvent, it is useful to make the generalization that like dissolves like, or a highly polar solvent such as water will tend to readily dissolve ionic or polar solutes, whereas nonpolar or only slightly polar solutes tend to be insoluble.

When water liquid with appreciable amounts of dissolved material is cooled, the point of maximum density is not at 4°C but is rather at the point of lowest temperature, regardless of the temperature that the solution attains below 0°C. This is because of the tendency of the ice crystal to exclude foreign nonwater molecules of dissolved particles from its crystalline structure.

Ice has a considerable amount of space between the water molecules that comprise the individual ice crystals. Regardless of the open structure of the ice crystal, foreign particles are not able to fit within the interstices between adjacent water molecules. As ice forms, these dissolved particles $Cl-$ and $Na+$, for example, are eliminated from the ice crystal in a random manner. Some of these particles are forced to the surface, where they form a salt film on the ice surface; while others are forced down into the unfrozen water, where they remain in solution. This increases the concentration of the particles that are dissolved in the unfrozen water. As the concentration of the solute increases, the mass of the solution also increases. As ice formation continues, the volume of the liquid water must decrease, since water is leaving the liquid state and entering the solid state (forming ice). The volume of this solution is, therefore, decreased as the mass is increased. This results in increased density of the solution.

In solute-solvent systems, the colder the water, the greater amount of ice that is formed. As ice formation increases, a greater number of dissolved particles will be excluded from the ice crystal and forced into the liquid water, while the volume of the liquid is reduced. Effectively increased is the density of the solution. In such systems, therefore, the point of maximum density is not at 4°C but is governed by the dissolved particle concentration of the remaining unfrozen liquid.

Freezing out of salts or crystallization also plays an important role in decreasing the freezing point of solutions. Pure water freezes at 0°C. However, if any given substance such as salt is dissolved in the water and the solution is then frozen, it will have a freezing point below 0°C. When a specific number of particles are dissolved in 1 kilogram (kg) of water or any liquid, they lower the freezing point of the solution by a predictable amount. This reduction is termed the *molal freezing-point depression*.

The common term, the *mole*, refers to the gram atomic weight of any element expressed in grams. It can be determined from the Periodic Chart that hydrogen has a gram atomic weight of 1.008. Therefore, 1 mole of H would weigh 1.008 g and would contain 6.02×10^{23} hydrogen atoms. One mole of chlorine (gram atomic weight = 35.453) would weigh 35.453 g and contain 6.02×10^{23} chlorine atoms. Although the weights change, the number of atoms contained in a mole will remain the same

When a given quantity of particles is dissolved in 1 kg of water, the freezing point of the solution will be lowered by a quantitative amount. When 1 mole of particles is dissolved in 1 kg of water, the freezing point will be decreased to minus 1.86°C; or the freezing point will be lowered below 0°C. It makes no difference whether it is 1 mole of Cl, 1 mole of Na, or 1 mole of Al that is dissolved. In each case, with 1 mole of solute dissolved in 1 g of water, the freezing point will be decreased by 1.86°C. This is known as the *molal freezing-point depression constant* of water. When predicting the freezing points of solutions that contain dissolved molecules, it must be considered as the process of dissociation when NaCl dissolves; it dissociates into Na+ and Cl- ions. Thus, if 1 mole (58.442 g) of NaCl were dissolved in 1 kg of water, the resulting solution would freeze at minus 3.72°C. This is due to the fact that when NaCl is dissolved in a polar solvent, it dissociates into its component parts. One mole of NaCl will dissociate into 1 mole of Na+ ions and 1 mole of Cl- ions. Consequently, the resulting solution will contain 2 moles of particles, and the freezing-point depression will be doubled (2×1.86°C). A polar molecule such as $CaCl_2$ when dissolved will dissociate into one Ca^{+2} ion and two Cl- ions. Therefore, 1 mole of $CaCl_2$ molecules dissolved in 1 kg of water will depress the freezing point to minus 5.58°C, since there are 3 moles of ions in the resulting solution.

Freezing-point depressions in solutions occur because the solute actually dilutes the solvent. The physical presence of the solute forces the solvent molecules further apart to provide space for the solute. In order to freeze, the solvent molecules must come into close proximity with each other to allow the hydrogen bonds needed to produce the crystalline structure to form. In order for this to occur, the solvent molecules must be made to move at a lesser rate than normal. Consequently, additional heat must be removed from the system to bring about the desired decrease in molecular motion. In other words, the temperature must be lowered below the normal freezing point. As a solution freezes, the unfrozen liquid portion increases in density due to the freezing out of the salts, but the freezing point is also depressed. Consequently, the colder the water gets, the denser it becomes, and the harder it is to freeze the remaining liquid. Eventually, a point is reached where it is impossible to attain the very low-pressure temperatures necessary to freeze extremely salty water (for example, marine systems). At this point, water will remain as a very dense, highly salty liquid water mass.

Gas Solubility

Boiling water or increasing the temperature will tend to lower the concentration of dissolved gases. When pure water is heated, more and more water molecules gain sufficient energy to allow them to escape from the liquid state. Similarly, if a beaker of water containing a dissolved gas is heated, the solute gas will gain energy and eventually attain a sufficient velocity to enable it to overcome the attractive forces and escape from the surface of the liquid, thus decreasing the concentration of the dissolved gas in solution. The reverse will occur if the water is cooled. In general, a given volume of cold water will contain a greater concentration of dissolved gas than an equal volume of warm water. Also, nonpolar gases such as O_2 are less soluble in water than polar gases such as CO_2.

Specific Heat

Of all the elements or compounds, only lithium, ammonia, and liquid hydrogen are capable of storing a greater quantity of heat with a smaller temperature rise. Because of water's heat-storing capacity it is recognized as having a high specific heat. Since very large quantities of heat are involved in aquatic temperature changes, both marine and aquatic, natural systems warm up slowly in the spring and cool slowly in the fall. This prevents wide seasonal temperature fluctuations and moderating terrestrial temperatures.

Latent Heat

Water has the highest heat of fusion and heat of evaporation or latent heat of any naturally occurring liquid. Heat of fusion is the large quantities of heat that must be removed to form ice and, conversely, the large quantities necessary to melt the ice. It takes 80 calories of heat to freeze 1 g of water--that is the equivalent to the number of calories required to raise the same quantity of liquid water at 0°C to 80°C. A body of water at 0°C can lose sufficient heat to warm a large amount of cold air with only minimal ice formation.

For heat of evaporation, the calories needed to convert 1 g of liquid water to vapor at a temperature of 100°C. This is equivalent to the amount of heat required to raise 536 g of water 1°C and is primarily due to the polar nature of water. The heat absorbed in evaporation is released in condensation and plays a major role in worldwide temperature changes.

3 CHEMICAL PROPERTIES

Water is exceedingly important in reactions of everyday concern, sometimes as a medium in which reacting substances are brought in contact with each other, sometimes directly with reactions themselves. Although the involvement of water in chemical reactions is sometimes a nuisance, with the widespread occurrence of water, such contacts are unavoidable. Reactions with water are also a result of many industrial processes.

The reactions that can be expected to occur when metal substances come in contact with water are considerable. All metals above hydrogen in the order of metal activity (electrochemical series) react with water more or less readily, whereas metals below hydrogen do not react, even under the most favorable conditions.

						Base or	Metal oxide
Water	+	Metal (above hydrogen)	→	Hydrogen	+	at ordinary temperatures	at elevated temperatures

Water with metal oxides are sometimes called basic oxides because they react with water to make bases.

$$\text{Water} + \text{Metal oxide} \rightarrow \text{Base}$$

One such reaction is the reaction of calcium oxide with water to make calcium hydroxide.

$$\text{Water} + \text{Calcium oxide} \rightarrow \text{Calcium hydroxide}$$
$$H_2O + CaO \rightarrow Ca(OH)_2$$

Such reactions range in extent from being quite marked with a soluble metal oxide to negligible with an insoluble oxide. The following are examples:

Water + Sodium oxide →(extensive reaction)→ Sodium hydroxide
(very soluble)

Water + Magnesium oxide →(slight but appreciable)→ Magnesium hydroxide
(slightly soluble)

Water + Copper oxide ⟶ Copper hydroxide (in negligible quantity)
(negligibly soluble)

Specific examples commonly encountered in everyday use are: water with iron oxide of iron rust, as found on rusty iron utensils, implements, or pipe and water with aluminum oxide of dulled aluminum ware. But because of the insolubility or negligible solubility of these metal oxides, reactions with water are very slight.

Water + Ferric oxide ⟶ Ferric hydroxide
(negligibly soluble)   (in negligible quantity)

Water + Aluminum oxide ⟶ Aluminum hydroxide
(negligibly soluble)   (in negligible quantity)

Water reacts with nonmetal oxides to produce acids. Nonmetal oxides are sometimes called acidic oxides.

Water + Nonmetal oxides ⟶ Acids

One such reaction is that between water and carbon dioxide to make carbonic acid.

Water + Carbon dioxide ⟶ Carbonic acid
H_2O + CO_2 ⟶ H_2CO_3

Since carbon dioxide is always present in the atmosphere, portions of this gas entering into solution react with some of the water. Thus streams, rivers, ponds, and lakes are very dilute solutions of carbonic acid. This reaction is of significance physiologically, since it takes place continuously in the body cells, and the carbonic acid thus made within the cells themselves or in the blood is a physiologically important factor.

Typical reactions of this type that are of interest include:

Water + Sulfur dioxide ⟶ Sulfurous acid
Water + Sulfur trioxide ⟶ Sulfuric acid
Water + Phosphorus pentoxide ⟶ Phosphoric acid

The reaction between water and sulfur dioxide is of importance industrially because it is one of the means that sulfurous acid is obtained for use as a bleaching agent and as a preservative. The reaction between water and sulfur trioxide is of consequence because it is necessary for making sulfuric acid by the contact process. Because of the tendency of water to react with phosphorus pentoxide, this oxide is one of the drying agents used in laboratories for removing water from moisture-containing gases to produce anhydrous gases.

HYDRATE FORMATION

Water reacts with certain compounds, many of which are salts, to make what are known as hydrates.

$$\text{Water} + \text{Certain compounds} \longrightarrow \text{Hydrates}$$

In hydrates water appears to be held in loose combination with the associated compound because of the ease with which hydrates give up water on heating or in a dry atmosphere. The water in combination within a hydrate is always in definite proportion by weight. These facts are illustrated in the formulas for hydrates; copper sulfate hydrate, formula $CuSO_4 \cdot 5\,H_2O$, is a typical example. Table 3-1 lists some commonly occurring hydrates.

Table 3-1 Commonly Occurring Hydrates

Hydrates of Common Occurrence	Common Names	Formula for Molecular Weight and Molecule
Calcium chloride hydrate		$CaCl_2 \cdot 6\,H_2O$
Calcium sulfate hydrate	Plaster of Paris	$(CaSO_4)_2 \cdot H_2O$
Calcium sulfate hydrate	Gypsum	$CaSO_4 \cdot 2\,H_2O$
Copper sulfate hydrate	Blue vitriol or bluestone	$CuSO_4 \cdot 5\,H_2O$
Ferrous sulfate hydrate	Copperas	$FeSO_4 \cdot 7\,H_2O$
Sodium borate hydrate	Borax (hydrate)	$Na_2B_4O_7 \cdot 10\,H_2O$
Sodium carbonate hydrate	Washing soda (hydrate)	$Na_2CO_3 \cdot 10\,H_2O$
Sodium sulfate hydrate	Glauber's salt	$Na_2SO_4 \cdot 10H_2O$
Magnesium sulfate hydrate	Epsom salt (hydrate)	$MgSO_4 \cdot 7\,H_2O$
Potassium aluminum sulfate hydrate	Alum	$KAl(SO_4)_2 \cdot 12\,H_2O$
Hydrogen sulfate hydrate		$H_2SO_4 \cdot 2\,H_2O$

Hydrates usually form crystals and sometimes the complex compound exhibits a color which is distinctly different from that of the anhydrous substance. This is shown in the behavior of copper sulfate. Anhydrous copper sulfate is a light gray powder; on contact with water a decidedly blue hydrate appears. If water is evaporated at ambient temperature from a solution of copper sulfate, the blue crystals of hydrate appear. The reaction for the equation is as follows:

$$\text{Copper sulfate} + \text{Water} \longrightarrow \text{Copper sulfate hydrate}$$
$$CuSO_4 + 5\ H_2O \longrightarrow CuSO_4 \cdot 5\ H_2O$$

When an anhydrous salt reacts with water to make a hydrate, there is always an energy product. This indicates that the hydrate possesses less chemical energy than did the anhydrous compound and water combined.

Reactions are reversible. When heat is supplied the hydrate decomposes to make the anhydrous compound and water. For example:

$$\text{Copper sulfate hydrate} \xrightarrow{\Delta} \text{Copper sulfate} + \text{Water}$$

Salts are not the only substances which react with water to make hydrates. Another example is sulfuric acid reacts in this manner. This is why concentrated sulfuric acid can be used as a drying agent for gaseous products of reactions that contain undesired moisture, since the concentrated acid combines with this water, thus removing it most effectively.

$$\text{Hydrogen sulfate} + \text{Water} \longrightarrow \text{Hydrogen sulfate hydrate}$$
$$\text{(Sulfuric acid)}$$
$$H_2SO_4 + 2\ H_2O \longrightarrow H_2SO_4 \cdot 2\ H_2O$$

Because the combination of sulfuric acid with water is accompanied by an unusually large quantity of heat product, bringing these substances together constitutes a hazard. When making a dilute solution of this acid from the concentrated supply, the laboratory worker should habitually pour the denser concentrated acid very slowly into the less dense water so that automatically a greater distribution of acid and water is obtained, and the heat product is more evenly distributed.

Hydrolysis is a double-decomposition reaction involving water as one of the reacting substances. In these reactions water functions as hydrogen hydroxide (HOH) with half of its hydrogen constituting a hydrogen radical and the rest of the hydrogen combines with all of the oxygen serving as a hydroxyl radical.

Many salts react with water, in double decomposition, to make an acid and a base. The exceptions are salts of strong acids and strong bases.

$$\text{Water} + \text{Sodium carbonate} \longrightarrow \text{(certain) Salts} \longrightarrow \text{Acid} + \text{Base}$$
$$\text{Water} + \text{Sodium carbonate} \longrightarrow \text{Carbonic acid} + \text{Sodium hydroxide}$$
$$2\ HOH + Na_2CO_3 \longrightarrow H_2CO_3 \longrightarrow + 2\ NaOH$$
$$\text{Water} + \text{Ferric chloride} \longrightarrow \text{Hydrochloric acid} + \text{Ferric hydroxide}$$
$$3\ HOH + FeCl_3 \longrightarrow 3\ HCl + Fe(OH)_3$$

Such reactions are not completed because the acid and base in the solution set up the reverse reaction to make water and salt. Thus, two opposing reactions are established in the solution.

$$\text{Water (in hydrolysis)} + \text{Salt} \rightleftarrows \text{Acid} + \text{Base}$$

There are always four substances in the solution: water, salt, acid, and base.

Relative Concentrations

The acid and base have a greater tendency to react to make water and salt than salt and water have to make acid and base. Consequently, the salt and water are always in relatively high concentrations and the concentrations of acid and base are relatively low in these solutions.

An example is a solution of sodium carbonate where only a small portion of this salt seemingly becomes hydrolyzed. The solution will contain water, the small portion involved in the reaction and the large portion that serves as solvent; most of the sodium carbonate, a little sodium hydroxide, and a little carbonic acid. The same conditions hold within a solution of the salt ferric chloride.

When a salt is composed of a metal radical from a strong base and an acid radical from a weak acid, it is referred to as a salt of a strong base and a weak acid. All salts of this type are subject to hydrolysis. Sodium carbonate, sodium bicarbonate, and potassium acetate are examples of these salts.

$$\text{Water (in hydrolysis)} + \text{Salt} \longrightarrow \text{Strong base} + \text{Weak acid}$$

When the water solution of such a salt is tested with litmus, the solution will be found to be basic reaction. In spite of the presence of both salt and weak acid, the solution behaves chemically like a dilute solution of the base.

When the metal radical of a salt is from a weak base, or if it is the ammonium radical, and the acid radical from a strong acid, the salt is referred to as a salt of a weak base and a strong acid. Salts of this type react with water in hydrolysis.

$$\text{Water (in hydrolysis)} + \text{Salt} \longrightarrow \text{Strong acid} + \text{Weak base}$$

Examples of these salts are ammonium chloride, copper sulfate, and ferric chloride. Representative of their hydrolysis reactions are as follows.

Water + Ammonium ⟶ Ammonium + Hydrochloric acid
(in hydrolysis) chloride   hydroxide

$$HOH + NH_4OCl \longrightarrow NH_4OH + HCl$$

Water (in hydrolysis)	+	Copper sulfate	⟶	Copper hydroxide	+ Sulfuric acid
2 HOH	+	$CuSO_4$	⟶	$Cu(OH)_2$	+ H_2SO_4
Water (in hydrolysis)	+	Ferric chloride	⟶	Ferric hydroxide	+ Hydrochloric acid
3 HOH	+	$FeCl_3$	⟶	$Fe(OH)_3$	+ 3 HCl

In the water solution of a salt of this type, a base and an acid are both present—as are also a salt and water. The solution, as a whole, is acid in reaction.

Water reacts in double decomposition with many substances which are not salts, and such reactions also are referred to as hydrolytic in character.

SOLUTIONS

When brought in contact with water or other selected liquids, substances dissolve either completely or in part forming solutions. The process of dissolving is automatic and proceeds in some instances rapidly, in other instances very slowly, until an even distribution is effected between the liquid and the dispersed substances. This condition is known as equilibrium. As a result, every solution eventually becomes homogeneous; that is, the solute or dispersed substance becomes evenly distributed throughout the solvent as the dissolving medium. Once the solution is obtained, no matter how long it may stand, provided the same temperature is maintained and no solvent is allowed to evaporate, no tendency toward separation of the solute from solvent becomes apparent even though there may be differences in their density. Other facts concerning solutions are that both liquid and dissolved solutes are able to pass through a filter, and they can diffuse through appropriate membranes. The dispersed particles are of such small dimensions that they cannot be detected even with the assistance of an ultramicroscope. The dispersed particles within a true solution are exceedingly small. In fact, their size has been computed to be less than 1 millimicron (1 mμ) in diameter, [1 micron (μ) = 0.001 mm millimicron (mμ) = 0.000001 mm] which means that they are probably single molecules of the solutes concerned, or ion particles.

Why solvent and solute portions of a true solution effect and maintain an even distribution regardless of the difference in their densities is not known. It is believed that within a true solution the dispersed particles of both solute and solvent are single molecules of the substances involved. These molecules are moving about with speeds which vary inversely with their relative weights; that is, the heavier the molecules the less the speeds they travel, while the lighter the molecules the greater their speed of movement. Despite the difference in relative weights of the molecules of solvent and solute, they all move about among each other with the same kinetic energy. There are frequent collisions among the molecules and against sides of the container followed by rebounds. In an attempt to reduce collisions to a minimum, the heavier but slowly moving molecules and the lighter but more rapidly moving molecules eventually effect an even distribution.

In many instances there is a tendency for molecules of water or other solvents to get in between and surround dispersed molecules of solute, thereby aiding in maintaining their distribution. Considering the character of the dispersion, a true solution is a system in which a substance is in even distribution throughout a liquid, with no tendency toward separation of the dispersed substance from the dispersing liquids.

A solution is a system in which the dispersed particles are single molecules of the solute in even distribution throughout the liquid solvent.

The dispersing liquid is commonly called the solvent and the dispersed substance the solute. Most frequently, the solute is a solid substance, although it can be a liquid, as is true of a solution of alcohol in water; or a gaseous substance, as in the case of a solution of hydrogen chloride in water, or oxygen in water.

In contrast to a true solution, a suspension is a dispersion system which can be obtained only through mechanical aids such as by stirring or shaking. A suspension is characterized by a tendency toward a separation of the dispersed particles from the dispersion medium. When mixing stops, separation sets in almost immediately. Examples of such systems are sand with water or raw starch with water. Dispersed particles which are characteristic of suspensions are greater than 0.1 micron (0.1 μ) in diameter or more than 100 times the diameter of the dispersed particles of true solutions. They typically cannot pass through a filter, nor are they diffusible through membranes and they can be distinguished either by the naked eye or with the aid of an ordinary microscope. Dispersed particles of a suspension are believed to be made up of many thousands of coherent molecules of the substances concerned.

Dispersions of a third type which range between true solutions and suspensions are called colloidal solutions or colloidal dispersions. An example of such dispersions is soap with water. In dispersions of this type the sizes of dispersed particles vary from 1 mμ in diameter, or larger than those of true solutions, to less than 0.1 μ diameter, or smaller than those of suspensions. Colloidally dispersed particles are presumed to be aggregates of many coherent molecules, the number of which can vary from less than 100 molecules in some aggregates to thousands in other aggregates; possibly in some instances colloidal particles may be extremely large, single molecules. The properties of colloidal solutions range between those of true solutions and those of suspensions. There is always a tendency for colloidal particles to separate from the dispersing liquid. If the aggregates are composed of relatively few molecules, the separation may not become apparent for a long time. Composed of thousands of coherent molecules, the particles will approach more nearly those within suspensions, and their separation from the dispersions will become apparent within a shorter period of time. Colloidal particles can pass through filters more or less readily, depending upon their size and as a rule they do not diffuse through membranes. They are not visible through an ordinary microscope but are present as discrete particles scattered through the dispersion which can be made evident through the aid of the ultramicroscope. Table 3-2 describes various properties of solutions and suspensions.

Table 3-2 Properties of Solutions and Suspensions

True Solutions	Colloidal Solutions	Suspensions
Less than 1 mμ in diameter	1 mμ to 0.1 μ in diameter	Greater than 0.1 μ in diameter
Indistinguishable even under ultramicroscope	Not visible under an ordinary microscope but presence detected by ultramicroscope	Distinguished by with the naked eye or with the aid of an ordinary microscope
Pass through filter paper, subject to diffusion through membranes	Pass through filter paper but usually do not diffuse through membranes	Does not pass through filter paper and not subject to diffusion
Dispersion is permanent	Tendency toward separation of dispersed substance more or less delayed	Separation of dispersed material is almost immediate

The same condition holds true for all other nonvolatile substances in water solution, with the exception of electrolytes, where the dispersed particles are single molecules. In general, then, for every gram molecular weight (or mole) of such nonelectrolytes dispersed in a liter of water, the boiling point of the solution will be increased 0.52°C over the boiling point of water alone, and the freezing point of the solution will be decreased 1.86°C below that of water. Thus, a solution that contains 2 gram-molecular weights of such a substance dissolved in a liter of water has a boiling point of 2 × 0.52°C higher than 100°C and a freezing point of 2 × 1.86°C lower than 0°C (the freezing point of water.)

Electrolytes are substances whose water solutions carry an electric current. Acids, bases, and salts are electrolytes. Some electrolytes enter a water solution as molecules, part of which separate into positively charged and negatively charged ions. In other instances, the electrolyte is an ionic compound which does not form molecules but which provides positively charged and negatively charged ions in the solution. Each ion and each molecule constitute a particle in the solution. It follows, therefore, that a solution of an electrolyte which contains a mole of the compound in a liter of water contains more individual particles than does a solution of a nonelectrolyte of similar concentration. Consequently, such electrolyte solutions show relatively higher boiling points and freezing points than do solutions of nonelectrolyte composition of like concentration.

For example, the salt, sodium chloride, is an electrolyte. The gram-molecular weight of sodium chloride is 58.448 gm. When this quantity of this salt is dissolved in 1 liter of water, the boiling point of the solution is 100.92°C and the freezing point is minus 7.72°C. Almost twice as great an effect is produced on the solution as is obtained by dissolving 1 mole of glucose which is a nonelectrolyte.

Osmotic Pressure

In a true solution molecules of water and those of the solute continually move about among each other in an attempt to effect and maintain an even distribution. In so doing they are continuously impacting against each other and against the sides of the container, or any other object in their way, and rebounding and changing the direction of their movement. An inner-solution pressure thus becomes established, which is called osmotic pressure, and is proportional to the concentration of dispersed particles of solute. The higher the concentration of the solute the greater the number of impacts involving solute and water molecules, hence, the greater the osmotic pressure of the solution. Conversely, the lower the concentration of the solute the less is this pressure within the solution.

An important phenomenon of osmotic pressure is that of osmosis, which is the diffusion of molecules of water or of water and solute through a permeable membrane in the attempt to equalize the solution concentrations on both sides of the membrane. The number of molecules of solute and of water that succeed in penetrating the membrane from either side will be in direct proportion to the number of their respective impacts against it.

The Universal Solvent

Every liquid is a solvent for selected substances. Water is the best known because it dissolves more substances to an appreciable extent than does any other liquid. Strictly speaking, no substance is absolutely insoluble in water, insoluble being merely a way of saying soluble only to an exceedingly slight degree.

There is no limit to the solubility of some substances in selected solvents. Examples are ethyl alcohol with water and of glycerol with water, which are soluble in each other in all proportions. But as a rule, a given quantity of water or of any other liquid solvent is able to hold in solution varying weights of a given substance up to a definite limit under given temperature conditions.

Saturated Solutions

A solution is saturated if upon contact with some of the undissolved solute, it is unable to take more into solution. For example, a solution of sucrose (203.9 gm in 100 ml of water) is saturated at the given temperature of 20°C because if in contact with an undissolved portion of this sugar, the quantity in solution does not increase. But when the solution is at 100°C, it requires 487.0 gm of sucrose in 100 ml of water to achieve saturation. Quantities of other substances that provide a saturated condition in the specified volume of water at given temperatures are listed in Table 3-3. Note that gaseous solubility decreases with increasing temperatures.

A solution is unsaturated when upon contact with some of the undissolved solute it is able to take more of the same into solution. For example, a solution of sodium chloride which at 20°C contains 30 gm of this solute in 100 ml of water is unsaturated because if some of the solid sodium chloride is introduced more will dissolve. Furthermore, the solution will continue to be unsaturated with this salt at this temperature until it contains 36.0 gm, at which time it will be saturated (see Table 3-3).

The solubilities of substances in water—meaning the quantities necessary to produce saturation—vary with the substances concerned and with the temperatures of the solution. *Slightly soluble* and *very soluble* substances are frequently used expressions, and with Table 3-3 at hand, they have significance. Mercurous chloride (calomel) is only slightly soluble in water at 20°C, which is around ordinary temperatures. On the other hand, silver nitrate is comparatively very soluble at ordinary as well as at high temperatures.

As a rule, the solubility of substances which are solids increases with an increase in the temperature of the solution; examples are given in Table 3-3. But there are a few solid substances that increase in solubility with decreasing solution temperatures; calcium hydroxide and calcium sulfate are more soluble in cold than in warm or hot water.

If the solutes are gaseous substances, they are always less soluble as the temperature of the solution increases, and vice versa. The speed with which a substance dissolves also is affected by both the area of contact between water and solute and the temperature of the solution. The greater the area of contact between water and solute, the greater the speed with which the substance dissolves. For this reason finely divided solids dissolve more rapidly than do the same substances in coarser particles. Stirring also increases solution by producing a thorough distribution of the solute in contact with water.

The effect of temperature in bringing about more rapid solution depends upon whether the solubility of the substance increases with increased temperatures or with decreased temperatures. For example, boric acid dissolves more quickly in hot than in cold water because its solubility increases with increasing temperatures of the solution. Calcium hydroxide dissolves more rapidly in cold rather than in hot water because its solubility increases with decreasing temperature of the solution. No general rule can be set forth as to the temperature most desirable for making solutions as rapidly as possible, but tables of solubilities of substances at different temperatures are helpful in this connection and should be consulted. The dissolving in water of some substances is accompanied by the appearance of heat, which is called the heat of solution of the substance. In such cases it is not wise

Table 3-3 Solubilities of Various Substances in Water

Solid Substances	Quantity of Substance in Grams to Produce Saturation in 100 mL of Water at the Given Temperature	
	at 20° C	at 100° C
Aluminum chloride	69.9 (at 15° C)	
Aluminum hydroxide	0.00015	
Aluminum oxide	Insoluble	Insoluble
Barium sulfate	0.00024	0.00039
Boric Acid	4.8	28.7
Calcium carbonate	0.0014 (at 25° C)	0.0018 (at 75° C)
Calcium chloride	59.5 (at 0° C)	159.0
Calcium hydroxide	0.165	0.077
Calcium oxalate	0.00068 (at 25° C)	0.0014 (at 95° C)
Calcium phosphate	0.002	
Calcium sulfate	0.1759 (at 0° C)	0.1619
Cupric oxide	Insoluble	Insoluble
Cupric sulfate hydrate	31.6 (at 0° C)	203.3
Glucose (dextrose)	83	
Iodine	0.029	0.078 (at 50° C)
Lactose (milk sugar)	17	
Magnesium oxide	0.0086 (at 30° C)	
Magnesium sulfate	26 (at 0° C)	
Mercuric chloride	6.1	61.3
Mercurous chloride	0.5	25.0
Potassium hydroxide	112.0	178.0
Potassium iodide	144.0	208.0
Silicon dioxide	Insoluble	Insoluble
Silver chloride	0.00089 (at 10° C)	0.0021
Silver nitrate	222.0	952.0
Sodium bicarbonate	9.6	16.4 (at 60° C)
Sodium chloride	36.0	39.8
Sodium hydroxide	109.0	347.0
Sodium tetraborate (borax)	2.83 (at 0° C)	201.4
Sucrose (cane sugar)	203.9	487.0
Zinc oxide	0.00016 (at 30° C)	
Zinc stearate	Insoluble	Insoluble

Gaseous Substances	In 100 mL of Water at Given Temperature Under Standard Pressure	
Carbon dioxide	179.7 ml at 0° C	90.1 ml at 20° C
Chlorine	310.0 ml at 10° C	177.0 ml at 30° C
Hydrogen	1.93 ml at 0° C	0.85 ml at 80° C
Hydrogen chloride	82.3 gm at 0° C	56.1 gm at 60° C
Oxygen	4.89 ml at 0° C	2.61 ml at 30° C
Nitrogen	2.35 ml at 0° C	1.55 ml at 20° C

to use hot water when making the solution concerned. These substances include sodium and potassium hydroxides, calcium and iron chlorides, iron sulfate, magnesium chloride, magnesium sulfate, and magnesium nitrate.

In some instances, when a solution has been saturated at a high temperature with a solid solute, the entire solute will remain in solution as the temperature of the solution is lowered, provided no undissolved solute is in contact with the solution. Such solutions at the lower temperature are supersaturated. A supersaturated solution is one which is holding in solution an excess of the solute above the quantity required for saturation at the given temperature.

Supersaturated solutions are frequently very difficult to maintain since a slight disturbance, such as shaking, stirring, or the introduction of a solid particle, especially if it is a particle of the solid solute itself (known as seeding), will give rise to a precipitation from the solution of the solid solute in excess of saturation at the given temperature. This is often the basis for crystallization.

Solutions are dilute or concentrated, depending upon the concentration of the solute, that is present in a given volume of the solution. Some substances are very soluble in water and therefore are capable of yielding highly concentrated solutions in this solvent.

Concentration Expressions

Various standards are used for solutions of more or less exact quantitative concentrations. Solutions are frequently made up on a percentage basis, and their concentrations are expressed as 1 percent solutions, 10 percent, 50 percent, and so on. If such solutions are accurately prepared, both the solute and solvent must be measured on a common weight basis. Thus, on a gram basis with water as solvent, a 10 percent solution of sodium chloride is one which contains 10 gm of sodium hydroxide to every 90 gm of water. For convenience, however, when water is the solvent used, although the solute may be measured in grams, the water is frequently measured in milliliters. For example, on this basis, a so-called 10 percent solution would contain 10 gm of solute to 90 ml of water. And a solution so assembled comes very close to being accurate on a percentage basis because, although actually 1 ml of water weighs 1 gm only when at 4°C, in the range of temperatures from 0°C to 100°C this volume (1 ml) of water always weighs very close to 1 gm.

Solutions on a percentage basis have no chemical significance, but they are appropriate for many desired purposes including industrial and laboratory uses. Solution concentrations which do have a chemical significance are called molar and normal solutions.

Molar solutions. One mole of a substance is 1 gram-molecular weight of the substance. With this information at hand we are in a position to define a molar solution as a solution which contains 1 mole of the solute in 1 liter of solution. A molar solution of sodium chloride contains 58.448 gm of this solute in 1 liter of solution. A molar solution of silver nitrate contains 169.888 gm of this compound in 1 liter of solution.

$$\underbrace{\underset{22.991}{Na} \quad \underset{35.457}{Cl}}_{58.448} \qquad \underbrace{\underset{107.88}{Ag} \quad \underset{14.008}{N} \quad \underset{3 \times 16}{O_3}}_{169.888}$$

More concentrated or more dilute solutions can be prepared on this basis as multiples or fractional parts of a molar solution. A 2-molar solution contains 2 moles or 2 gram-molecular weights of solute in each liter of solution, whereas a one-tenth-molar solution contains one tenth of the gram-molecular weight of the solute in each liter of solution.

A solution of molar concentration is represented by 1 M, a 2-molar solution by 2 M, a solution containing 5 moles of solute per liter of solution by 5 M. Similarly, for more dilute solutions, the concentration of a solution which contains one half mole of solute per liter is represented either by 0.5 M or by M/2, a one-tenth-molar solution is represented by 0.1 M or M/10, a one-fifth-molar solution by 0.2 M or M/5, etc. Examples of some of these solution concentrations on a molar basis are as follows:

	Contains
1 M (molar) solution Sodium chloride solution	58.448 gm in 1 liter of solution
2 M (2-molar) solution Sodium chloride solution	116.896 gm in 1 liter of solution
0.1 M (ore-tenth-molar) solution Sodium chloride solution	5.8448 gm in 1 liter of solution

Solution concentrations on this basis have a chemical significance in that they are made up on a molecular-weight basis, and molecular weights serve as measuring units for substances as they become involved in chemical reactions. If the ratio of molecular weights of substances mutually involved in a given chemical reaction are known and if solutions of the substances are made up on a molar basis, it is possible to measure out volumes of the solutions respectively concerned to contain the exact weights of the substances for their complete reaction.

For example, in the reaction between sulfuric acid and sodium hydroxide, the equation provides data as to the molecular-weight ratio that must always be met for exact reacting-weight quantities of these substances.

$$\text{Sulfuric acid} + \text{Sodium hydroxide} \rightarrow \text{Water} + \text{Sodium Sulfate}$$
$$H_2SO_4 + 2\,NaOH \rightarrow 2\,H_2O + Na_2SO_4$$

1 gram-molecular weight **Sulfuric acid**	2 gram-molecular weights **to**	**Sodium hydroxide**
1 liter of M solution	to	2 liters of M solution
or, 10 ml of M solution	to	20 ml of M solution
or, 10 ml of 0.1 M solution	to	20 ml of 0.1 M solution

Since 1 gram-molecular weight or 1 mole of every substance is believed to include 6.02×10^{23} molecules of the substance, it follows that 1 liter of a molar solution must contain 6.02×10^{23} molecules of the substance concerned, while 1 liter of a one-half-molar solution must contain one half this number of molecules of the substance. This means that equal volumes of solutions of various substances of the same molarity must contain equal numbers of molecules of these substance.

Normal solutions. Solutions on a normal basis apply only to acids, bases, and salts, and they are concerned with concentrations of the radicals provided within these compounds rather than with concentrations of the compounds in themselves. The measuring unit for the solute is the equivalent weight; that is, the weight of the compound which provides a 1-valent quantity of each of the constituent radicals. A normal solution contains 1 gram-equivalent weight of the acid, base, or salt in a liter of the solution. Table 3-4 provides examples.

Solutions of greater or lesser concentration but still on a normal basis are also frequently used. Shorthand representations for solution concentrations on a normal basis are as follows: N, 2 N, 5 N, obviously with respect to normal, 2-normal, and 5-normal concentrations; 1/2 N or 0.5 N and 1/10 N or 0.1 N with respect to solutions of one-half-normal and one-tenth-normal concentrations.

A normal solution of an acid or of a base may also be defined on the basis of the concentration of the hydrogen radical or of the hydroxyl radical, respectively, provided in the solution. Thus, a normal solution of an acid is a solution which contains a 1-valent quantity, or 1.008 gm, of hydrogen radical (H+) in 1 liter of solution; whereas a normal solution of a base is a solution which contains a 1-valent quantity, or 17.008 gm, of hydroxyl radical (OH-) in 1 liter of solution.

Modifications of normal solutions of acids and bases, such as one-half-normal and one-tenth-normal solutions, are of commensurate concentrations.

Milliequivalent weight is used frequently for recording the concentration of various elements or radicals contributing to physiological solutions, such as blood or urine. The milliequivalent is simply 1/1,000 of the corresponding gram-equivalent weight; or it may be interpreted as the equivalent weight of the element or radical recorded in milligrams, rather than in grams. For example, a solution containing 5 mg/liter of calcium (Ca^{++}) means that there are 100.2 mg of calcium radical per liter of solution.

COLLOIDAL SOLUTIONS

Properties of colloidal solutions in contrast with related properties of true solutions have been considered previously. The difference in properties of these two varieties of dispersions is due primarily to the difference in the sizes of their dispersed particles. Particles in true solutions are single molecules, with exceedingly minute dimensions; those in colloidal solutions are molecule aggregates which may vary from clusters of possibly less than 100 molecules to aggregates of thousands of coherent molecules.

Colloidal particles remain dispersed, and remain dispersed for a longer or shorter time among single, lightweight water molecules of the dispersion medium. Reasons for this phenomenon include:

Table 3-4 Normal Solutions

A Normal Solution of	Contains in 1 Liter of Solution
Hydrochloric acid H^+ Cl^- 1.008 35.457 ⸺⸺⸺⸺⸺ M. W. (36.465)	1 gram-molecular weight (36.465 gm) of this acid, since this provides for the presence in the solution of a 1-valent quantity of each of the two radicals: 1.008 gm hydrogen radical (H^+) 35.457 gm chloride radical (Cl^-)
Calcium hydroxide Ca^{++} $(OH)_2$ 40.08 2 X 17.008 ⸺⸺⸺⸺⸺ M. W. (74.096)	1/2 the gram-molecular weight (37.048 gm) of this base because this contributes to this volume of solution a 1-valent quantity of each of the two radicals: 20.04 gm calcium radical ($Ca^{++}/2$) 17.008 gm hydroxyl radical (OH^-)
Sodium phosphate Na_3^+ $(PO_4)^{---}$ 3 X 22.991 94.975 M. W. (163.948)	1/3 the gram-molcular weight (54.657 gm) of this salt which contributes to 1 liter of solution a 1-valent quantity of each of the two radicals: 22.991 gm sodium radical (Na^+) 31.325 gm phosphate radical ($PO_4^{---}/3$)

- Water molecules, although light in weight, are in rapid vibratory motion, moving about in every direction. As heavy colloidal particles start downward, they are being continually hit by numerous water molecules which drive them first in one direction then another—down, up, sideways, and so on. As a result, colloidal particles follow a sort of random zigzag path that is referred to as Brownian movement. Obviously, the impacts by water molecules do not check entirely the precipitation of these particles but do cause more or less an indefinite delay.
- Colloidal particles themselves are in more or less vibratory motion which results in impacts against water molecules, against each other, and against the sides of the container, and these impacts are accompanied by reboundings into the dispersion.
- The most significant factor in the dispersion of colloidal particles is that they usually carry either positive or negative charges. As a result, a repulsive effect becomes established between these particles, and in their effort to keep away from each other they tend to remain distributed in the dispersion water medium.

Because of the weight of the colloidal particles, there is always a tendency toward precipitation despite the forces described which function to maintain their dispersion. Although this separation may not become apparent for long periods of time, the predisposition is present. Moreover, the heavier particles give evidence of precipitation sooner than lighter ones. A potent precipitating effect is produced by the introduction of an acid, base, or salt, the positive or negative ions from which offset opposite charges on the colloidal particles; therefore, with repulsive forces between these dispersed particles being diminished, precipitation begins.

True solutions with water as the solvent are transparent, since a light beam passes through them unimpeded by the tiny dispersed particles which are single molecules. Colloidal dispersions are from slightly to decidedly translucent. This is due to the fact that upon penetrating a colloidal solution, light rays strike against the sizable colloidal particles and are reflected at various angles, or scattered, throughout the dispersion. This effect is quite conspicuous in a dispersion of albumin in water and in a gelatin water dispersion. There are some colloidal dispersions which appear to the naked eye to be transparent but, when examined with the ultramicroscope, are shown to be colloidal in character. In such dispersions the colloidal particles are relatively small.

Substances in true solution have the effect of increasing the boiling point above that of water alone and depressing the freezing point below that of water. Substances in colloidal dispersion produce from very little to no appreciable effects on these properties, with the result that colloidal dispersions in water have boiling points of $100°C$, or very little above this temperature, and freeing points of $0°C$, or very little below this level. The reason is that these effects are proportional to particle concentrations, and in any colloidal dispersion, while there are many thousands—even millions—of molecules, they are clustered in relatively few molecule aggregates which are the dispersed particles.

Dispersed molecules of true solutions are able to diffuse through permeable membranes. The dispersed particles, or molecule aggregates, of colloidal solutions are incapable of diffusion through such membranes or can do so only to a very slight extent. More especially, it is the size of the particles that is involved. Most colloidal particles are too large to squeeze through the minute pores of the permeable membrane. Only those that are made up of relatively few coherent molecules are able to accomplish this penetration occasionally.

The difference in the diffusion possibility of substances in true solution and those in colloidal solution is taken advantage of for the separation of two such substances when they are mingled in the same water dispersion. This is achieved by enclosing the dispersion in a bag of permeable membrane which is immersed in water. The substances in true solution are able to pass through the membrane into the surrounding water, but the substances in colloidal solution are not. This process of separating a colloidal substance from a substance in true solution by diffusion of the soluble substance through a membrane is sometimes called dialysis.

This difference in diffusion possibilities of substances in true and colloidal solutions is of vital importance to plants and animals. The water of the sap of plants and of the blood of animals holds some substances in true solution and others in colloidal solution, but only the substances in true solution are able to enter the cells. On the other hand, within the cells of both plants and animals there are substances held in colloidal solution that must be kept there if cell activities are to proceed normally. Fortunately, being colloidal, these substances are incapable of escaping from the cells.

Colloidal dispersions are referred to as *sols*, meaning dispersions which are fluid. This is in contrast to dispersions called *gels*, which are jellylike in character. Although containing considerable water, gels possess a greater or lesser degree of rigidity or viscosity. A colloidal gel is possible with a substance which is capable of concentrating great quantities of water at the surface of its dispersed particles and in so doing swells enormously. Familiar examples of gels are fruit jellies and the more or less rigid dispersions made from gelatin. The term *gel* is derived from *gelatin*, which is capable of forming this variety of dispersion.

Gels may vary from fluid jellies, which flow under slight pressure, to gels with great rigidity. The rigidity is believed to be due to a development within the dispersed system of a mesh or sponge like structure that is brought about presumably by the adherence of some of the colloidal particles to one another in such a manner as to form a network of fibrils which encloses a more dilute colloidal solution.

EMULSIONS

Emulsions can be defined as a mixture of two mutually insoluble liquids, one of which is dispersed in the other in more or less finely divided droplets. The most commonly occurring emulsions are those of water with fatty oils and water with liquid hydrocarbons, or mixtures of liquid hydrocarbons such as gasoline, kerosene, and oil. Dispersed droplets in an emulsion are usually of suspension dimensions, hence, large enough to be discernible to the naked eye or under an ordinary microscope. Such a dispersion is a suspension emulsion—most French dressings are of this type. Occasionally, however, the dispersed droplets are colloidal in size in which case the dispersion is a colloidal emulsion. The more permanently dispersed systems of some mayonnaise dressings are examples.

While emulsions can be obtained easily by mixing and shaking the two liquids together, they are difficult to maintain unless some means is provided to check the tendency of the two liquids to separate. Such a means is provided through use of selected substances which, when added to the emulsion, are capable of prolonging or stabilizing the dispersion period. Such substances are known as emulsifying agents or stabilizers. The mixture of an oil with water, first without, then with the presence of a stabilizer, is an example. Upon thorough shaking of the oil with water, a suspension is formed but, if allowed to stand, within a very short time practically all of the oil will be found in a layer on top of the water. However, if a soap, surfactant, or wetting agent is added and the mixture is again thoroughly shaken, a more lasting emulsion is obtained with the wetting agent acting as an emulsifying agent. Emulsions are characterized by a milky appearance.

The functioning of the emulsifying agent in prolonging the life of an emulsion is due to its ability to collect in a thin film at the surface areas of the dispersed droplets, thereby preventing them from coalescing to make larger drops. In order that it may function in this way, the emulsifying agent must be a substance which can be concentrated or adsorbed either by the water or by the dispersed liquid at surface areas called interfacial areas where the water and dispersed substance meet. A factor in the behavior of emulsions and in the functioning of emulsifying agents is the surface tensions or interfacial tensions of the two participating liquids and the changes which are effected in these surface tensions by the introduction of an emulsifying agent.

IONS AND IONIZATION

Electrolytes and Nonelectrolytes

Acids, bases, and salts are the only compounds whose water solutions conduct an electric current. For this reason they are called electrolytes. This is in contrast to many other compounds, such as sugars and alcohols, which are called nonelectrolytes because their solutions do not conduct an electrical current.

Different electrolytes, even in solutions of comparable concentrations, differ as to the degree in which they conduct the current of electricity, and on this basis they are classed as strong electrolytes and weak electrolytes. It is significant, moreover, that strong electrolytes are very active chemically, whereas weak electrolytes are relatively less active chemically. Chemical activity or strength of a given acid, base, or salt is directly related to the degree to which this electrolyte, in solution, conducts the electric current.

A strong electrolyte may be described as an acid, base, or salt which is active chemically and, in water solution, gives good conduction of an electric current. By contrast, a weak electrolyte is an acid, base, or salt which is not very active chemically and which, in water solution, is a relatively poor conductor of electric current. An important observation regarding acids, bases, and salts is that to exhibit electrical conductivity and the associated chemical activity peculiar to these compounds, they must be in solution in water. Because of this and other phenomena which are peculiar to electrolytes, chemists long ago began to suggest possible explanations in terms of something which happens to acids, bases, and salts in water solution.

Arrhenius theory proposes that when acids, bases, and salts dissolve in water, part or all of their molecules separate to form independently reacting, electrically charged particles which have come to be called *ions* (meaning *wanderers*). For example:

$$NaCl \longrightarrow Na^+ + Cl^-$$
$$Ca(OH)_2 \longrightarrow Ca^{++} + 2\,(OH)^-$$
$$HC_2H_3O_2 \longrightarrow H^+ + C_2H_3O_2^-$$

These ions, rather than the neutral molecules, are the chemically active particles that engage in the reactions peculiar to acids, bases, and salts, and the conduction of the electric current is accounted for by the flow of these charged particles through the solution. The different degrees of activity exhibited by solutions of different electrolytes are explained, according to this theory, by the proportion of active ions to molecules in the solution. Thus, with a strong electrolyte, such as sodium chloride or potassium hydroxide, the proportion of ions in the solution is high—even up to 100 percent whereas in solutions of weak electrolytes, such as acetic acid and ammonium hydroxide, the neutral molecules predominate and the proportion of ions is correspondingly low.

4 THE EXAMINATION OF WATER

The examination of water is undertaken in the laboratory for many reasons. Probably one of the most important reasons is to form an opinion on the suitability of a water supply for use or treatment. This involves consideration of various factors such as whether the water is safe for human consumption, whether it is corrosive to metal pipes or will form scale, whether it is pleasant in appearance and taste, whether it is satisfactory for use in domestic washing, and whether it can be used for industrial purposes. Laboratory analyses are necessary for the control of water treatment to insure satisfactory water at all times. Various tests are tools which supplement and extend human senses. Gradual deterioration of various treatments such as coagulation and filtration processes can be detected by laboratory measurements before it becomes evident to visual observation; microscopic organisms can be magnified and counted, variation in dissolved constituents of water can be detected and unsatisfactory plant operation diagnosed so that corrective measures can be adopted.

The examination may be divided into physical, chemical, bacteriological, and microscopic categories. The physical tests give a measure of the properties detectible by the senses. Chemical analysis detects the presence of mineral and organic materials affecting the quality of the water, present as pollution and shows the effects of the treatment process—a necessary control measure. Bacteriological examinations indicate the presence of bacteria characteristic of sewage pollution and thus establish the safety of the water for consumption. Microscopic examinations provide information concerning growths of minute plants and animals in the water, frequently the cause of disagreeable tastes and odors and of filter clogging.

MEASUREMENT UNITS

In the laboratory examination of water, units of measurement are employed that are more convenient than those in everyday use. Most of the quantities measured are relatively small. Because fractions of the pound or quart are cumbersome, the metric system is widely used. The standard weight is the kilogram. Laboratory units most frequently used are the gram, equal to one thousandth of a kilogram;

and the milligram, one thousandth of a gram. The system is based on Latin prefixes *kilo* = 1,000; *deci* = 1/10; *centi* = 1/100; and *milli* = 1/1,000.

The unit of volume is the liter. For ordinary work the weight of 1 liter of water at room temperature may be considered to be 1 kilogram. This is exact, however, only when the water temperature is 4°C.

The unit of length is a meter which is divided into centimeters and millimeters, 1/100 and 1/1,000 of a meter, respectively.

Temperatures are measured in the Centigrade scale. The standard unit is 1/100 the difference in temperature between that of the melting point of ice and the boiling point of water at 760 millimeters atmosphere pressure. To convert Centigrade degrees to Fahrenheit degrees, the following equation is used:

$$\text{Degrees Centrigrade} = \frac{\text{degrees Fahreneit} - 32}{1.8}$$

The same equation can be manipulated to determine the degrees Fahrenheit when the Centigrade temperature is known.

$$\text{Degrees Fahrenheit} = \text{Degrees Centigrade} \times 1.8 + 32$$

or written in another way:

$$\frac{F-32}{180} = \frac{C}{100}$$

To report the small amounts of chemicals and other substances that are present in water without using fractional units, the ratio of parts per million (ppm) is used. In the metric system this is defined as 1 milligram per million milligrams, the weight of 1 liter of water.

$$1.0 \text{ ppm} = 1 \text{ mg per liter } (1,000,000 \text{ mg})$$

Using the English system of measurement, 1 pound per million pounds of water would be expressed as 1.0 ppm.

Since 1 gallon of water weighs 8.34 pounds, a million gallons would weigh 8,340,000 pounds. Thus, 8.34 pounds of a given substance in 1 million gallons of water would be expressed as 1.0 ppm.

$$\text{ppm} \times 8.34 = \text{pounds per million gallons}$$
$$1 \text{ ppm} = 8.34 \text{ pounds per million gallons}$$

The English units, grains per gallon, are often used, especially in reporting chemical doses. One pound equals 7,000 grains and 1 gpg = 17.1 ppm.

Milligrams per liter (mg/l) is generally used rather than parts per million (ppm). Parts per million expresses a weight/weight ratio, and water is almost never weighed for analysis. The unit is therefore inconsistent with practice, since almost always a result reported as ppm has actually been determined as mg/l, on a weight/volume basis. Milligrams per liter (mg/l) is the official terminology in *Standard Methods*.

Sampling

Laboratory tests depend upon sampling methods, and samples must be representative of the water to be examined or the results have no significance.

Representative Samples. A raw sample collected at the surface of a reservoir might have little relation to the water entering the intake. If a sample is collected from a tap, the water should first be run to waste long enough to empty the service pipe and thus obtain a sample representative of the water in the main. Normally, dead ends should be avoided. Samples collected after chlorination must be taken at a point where the chlorine has become completely mixed with the entire volume of water for a period of at least 10 minutes. Extraneous material must be excluded and the stopper or neck of the bottle should not become soiled. Samples of filtered water are generally taken from the clear water well. On occasion, separate samples from each filter may be required to detect faulty operation.

Sampling Frequency. Frequency of sampling must be determined for each supply and for each individual treatment plant. Characteristics of water from well supplies or large storage reservoirs fluctuate much less than those of water from small sources or streams. Water from lakes, either large or small, may fluctuate rapidly with changing winds or flow of tributary streams or weather. Semiannual samples from some wells may be sufficient, while monthly samples may be required from others. For surface water supplies weekly, and in many cases daily or even hourly, samples may be necessary to control treatment processes. The frequency of plant samples depends on fluctuations of the raw water quality. With a fairly constant raw water, weekly samples should suffice for most tests. For control of chlorination a minimum of daily and, more often, hourly samples may be required. For bacteriological examination, samples of raw and finished water should be examined daily where coagulation, filtration, and chlorination are practiced. For control of the sanitary quality of water in the distributing system, Table 4-1 may be used as a guide, although this is the minimum number of samples that is acceptable.

Bacteriologic Examination. Only clean, wide-mouthed, 4-ounce plastic or glass-stoppered bottles of Pyrex, Kimax, or Nosolvit glass should be used for samples to be examined bacteriologically. Ordinary glass will not stand the repeated heating needed for sterilization and may impart enough alkali to a sample to be bactericidal. The stopper, neck, and mouth of the bottle must be protected from contamination. This is generally accomplished by covering these areas with metal foil or heavy paper.

Table 4-1 Number of Samples Recommended Based on Population

Population served	Minimum number of samples per month
2,500 and under	1
10,000	7
25,000	25
100,000	100
1,000,000	300
2,000,000	390
5,000,000	500

Samples for bacterial examination always should be collected in sterilized bottles, since contaminated bottles will prevent accurate evaluation of the results. Sterilization can be accomplished by heating the bottles at 170°C for 1 hour, timing to start after the temperature in the oven has reached 170°C. Autoclaving at 121°C for 30 minutes is also acceptable. When samples of chlorinated water are to be examined, the residual chlorine must be destroyed when the sample is collected or the results will not be typical of the water at the point of collection. The presence of a dechlorinating agent in the sample bottle will neutralize any residual chlorine and will prevent continuation of the bactericidal action during the time the sample is in transit to the laboratory. The addition of 0.1 ml of a 10 percent solution of sodium thiosulfate ($Na_2SO_3 \cdot 5H_2O$) to the bottle before sterilization will provide for neutralization of 15 ppm chlorine in a 100-ml sample.

In the sampling procedure, sterilized bottles for bacteriological samples should be handled with care to avoid contamination. Leaky taps should be avoided since water flowing over the surface of the tap would contaminate the sample. Holding the bottle at or near the bottom, loosen the string around the protective cap and remove the stopper with the cap in place. Discard the tagged string sometimes placed between the stopper and the neck of the bottle to prevent the stopper from sticking. Care must be taken that the exposed stopper does not touch anything that might contaminate it. The lip of the bottle must not be contaminated by the hands, and the water must not flow over the hands into the bottle. Fill the bottle to within half an inch of the stopper, leaving only sufficient air space for expansion. Replace the stopper and tighten the string securely around the protective cap.

Sample bottles for the collection of samples for chemical analyses should be clean but need not be sterile. For a complete chemical analysis, about one gallon of the sample is required, but for plant operation this is seldom necessary and a one- to two-liter sample is sufficient.

The sampling points at which samples should be taken depend upon the purpose of the examination and the need for obtaining samples representative of the water to be examined. Thus, it is not possible to specify sampling points in general that would be applicable to a particular water supply.

Samples of raw water should be collected to determine the characteristics of the water that are to be corrected by treatment. These characteristics fluctuate in varying degrees with different waters and have marked effects on plant operation. When coagulation and filtration are employed, samples of the coagulated water demonstrate the efficiency of such treatment as do samples of filter effluent. Samples of the water as it enters the distribution system show the overall efficiency of the treatment employed. Samples from the distribution system show the character of the water delivered to the consumer and, by comparison with the treated water, show if any changes occur during distribution.

Generally, a sample of water as it enters the distribution system and several samples from the distribution system are desirable for this purpose. The number taken from the system varies with the facilities available, the population served, and the purpose for which the examinations are made. Normally, samples for bacteriologic examination are collected from many scattered points on the system while samples for chemical analysis are collected from only a few points.

COLOR

The color of water is commonly caused by the extraction of coloring material from humus of forests or from vegetable matter in swamps and low-lying areas. This coloring matter is composed of humus and tannic acid compounds which cause the yellowish brown tea color of surface waters. The color of water is of two types: (1) *true color* is that present in the water after the suspended matter has been removed, and (2) *apparent color* is true color plus any other color produced by substances in suspension. In certain cases, color may be imparted to water by dissolved iron or by the discharge of industrial wastes, or by microscopic organisms.

Color has in general little relation to pollution except as indicating surface water reaching groundwater supplies and the presence of contaminants. The attractiveness of water depends on the color to which the public of any locality has become accustomed. The removal of color is a function of water treatment, and therefore, a decrease in color is a measure of plant efficiency. Natural color is due to a wide variety of substances, and it has been necessary to adopt an arbitrary standard for its measurement. The color produced by 1 mg/l of platinum in the form of potassium chloroplatinate (K_2PtCl_6) is taken as the standard unit of color. Determinations for color should be done on the raw water, coagulated and filtered water, and on water from the distribution system. Only clean glass bottles of at least 500-ml capacity should be used for collecting samples.

Method of Determination

- Equipment
- Clean glass sample bottles
- 6-inch glass funnel

- Funnel support
- Filter paper, Whatman #30, 180-mm diameter
- Pipette, Mohr measuring 10-ml capacity, graduated in 1/10 ml
- 18 color-comparison tubes, Nessler A.P.H.A. high-form 50 ml
- Color tube support for high-form Nessler tubes
- Reagents
    - 100 ml of platinum—cobalt color standard, 500 units
- Procedure
    - To prepare standards having color units of 5, 10, 15, 20, 25, 30, 35, 40, 50, 60, and 70, dilute 0.5, 1.0, 1.5 ml, and so on of the standard color solution with distilled water to 50 ml in Nessler tubes. These standards, if protected from dust and evaporation, are stable for six months.
    - Fill a Nessler tube to the 50-ml mark with sample and compare with standards by looking vertically downward through the tubes towards a white surface placed at such an angle that light is reflected upward through the columns of liquid. This is the apparent color.
    - To determine the true color if suspended matter or turbidity is present, filter the sample through filter paper (Whatman #40 or equal) and compare the filtrate in a 50-ml Nessler tube with standards as previously.
    - Waters having color greater than 70 units are diluted with distilled water before comparison.
    - As the standards are prepared directly in color units, the standard that matches the sample records the color. If the sample has been diluted, the reading must be multiplied by the amount of dilution.

$$\text{Color units} = \text{Reading} \times \text{Dilution factor}$$

- Interpretation

    The U.S. Public Health Service recommends that waters intended for human use should not have a color exceeding 15 units. Coagulation and filtration should reduce the color to less than 5 units. A gradual increase in color of filtered water indicates impairment of either coagulation or filtration efficiency. Slow sand filters generally remove about 40 percent of the color of the raw water. Increase in color between the clear well and the distribution system may be caused by corrosion or growths in the pipes.

ODOR

Odors in water are caused by extremely small concentrations of volatile compounds. Since odorous materials are detectable when present in only a few micrograms per liter, analytical procedures are unsatisfactory for their measurement and reliance is placed on the sense of smell. This varies with individuals and, therefore, the results also vary. Care must be taken to avoid fatigue, as the ability to detect slight odors is quickly lost when used for any length of time, or if strong odors are encountered. Of equal importance is the performance of the test in an odor-free room and with odor-free equipment.

Some odors are produced when organic matter decomposes and are likely to be present in surface waters due to the presence of organic matter from surface wash. Intensity and offensiveness vary with the type of organic material, some being earthy, greasy, and musty, while others are putrefactive. Industrial wastes such as phenolic or oil wastes are responsible for some of the odors in surface waters.

Most of the objectionable odors in surface waters are caused by plankton, which liberate minute traces of volatile essential oils. While the organisms are increasing in concentration, the odors are not so strong as when they are decreasing and the dead organisms are decomposing. The use of chlorine for disinfection frequently destroys odor-producing substances, but in many cases has also been known to accentuate specific types of odors.

Odor can be measured by the *threshold odor number* which is the degree of dilution of the sample with odor-free water required to reduce the odor to the point where it is just detectable. The formula for calculating the value is:

$$\text{Threshold odor value} = \frac{\text{Volume of sample} + \text{Volume of odor-free water}}{\text{Volume of sample}}$$

Odor also can be described according to the odor characteristics given in Table 4-2. The intensity of the odor may be designated by prefixing a numeral. Odor is generally determined only on the raw and finished water, although samples from the distribution system should be examined at frequent intervals. Samples should be taken with as little aeration as possible or odor will be lost, since the odorous material may be vaporized or destroyed by dissolved oxygen in the air. The bottle used for the test should be only about two-thirds full to allow space for shaking the sample. The method of determination consists of:

- Equipment
    - Clean, odor-free glass bottles. In most plants the use of a good detergent and rinsing in distilled water will remove odor from glassware.
    - Large hot plate or water bath.
- Procedure
    - Warm sample and bottle to room temperature.
    - Shake sample.
    - Remove stopper and sniff odor at mouth of bottle.
    - Record odor characteristic and intensity as given in Tables 4-2 and 4-3.

The U.S. Public Health Service recommends a threshold odor limit of 3 which should not be exceeded in potable waters. Relative to odor intensity as listed in Table 4-2, a delivered water should not exceed a value of 2 to be acceptable.

Table 4-2 Odor Intensity with Prefixed Quality

Numerical value	Term	Definition
0	None	No odor perceptible.
1	Very faint	An odor that would not be detected ordinarily by the average consumer, but that could be detected in the laboratory by an experienced observer.
2	Faint	An odor that the consumer might detect if his or her attention were called to it, but that would not attract attention otherwise.
3	Distinct	An odor that would be detected readily and that might cause the water to be regarded with disfavor.
4	Decided	An odor that would force itself upon the attention and that might make the water unpalatable.
5	Very strong	An odor of such intensity that the water would be absolutely unfit to drink (a term to be used only in extreme cases).

Many communities become accustomed to a particular odor and object strongly to any change in the character of such an odor. The character and intensity of odor often is an aid to the interpretation of other examinations and an indicator of the character of pollution. The test yields results that are necessary to evaluate the effectiveness of various types of water treatment, particularly where odor removal is a function of treatment.

Express the intensity of the odor by a numeral prefixed to the term expressing quality, which may be defined as shown in Table 4-2.

Table 4-3 lists various compounds and a description of their odors.

TASTE

Taste in water is generally closely related to odor and caused by the same conditions.

- Taste is a factor of consumer acceptance of water, although as with odor, a change in character may cause complaints if consumers have become accustomed to one type of taste.

Table 4-3 Odor Characteristics

Nature of Odor	Description—Such as Odors of
Aromatic (spicy)	Camphor, cloves, lavender, and lemon
Cucumber	Synura
Balsamic (flowery)	Geranium, violets, and vanilla
Geranium	Asterionella
Nasturtium	Aphanizomenon
Sweetish	Coelosphaerium
Violets	Mallomonas
Chemical	Industrial wastes or chemical treatment
Chlorinous	Free chlorine
Hydrocarbon	Oil refinery wastes
Medicinal	Phenol or iodoform
Sulfuretted	Hydrogen sulfide
Disagreeable	Pronounced unpleasant odors
Fishy	Uroglenopsis and Diobryon
Pigpen	Anabaena
Septic	Stale sewage
Earthy	Damp earth
Peaty	Peat
Grassy	Crushed grass
Musty	Decomposing straw
Moldy	Damp cellar
Vegetable	Root vegetables

- Dissolved mineral matter may impart tastes but no odor to water.
- Metallic salts such as copper, zinc, or iron may cause metallic tastes.
- Chlorides or sulfates above 250 mg/l make the water salty in taste.
- Chlorinated water containing phenolic compounds may have a distinct taste in concentrations below that detectable as odor.

The characteristics and intensity of taste are described as the same as for odor as listed in the preceding tables. Tests for taste are generally made using the same samples as collected for odor and determined as follows:

- Equipment
  - Clean, taste-free sample bottles
  - One 50-ml beaker
- Procedure (Water should not be tasted if there is any doubt regarding its sanitary quality.)
  - Warm sample to room temperature.

- Pour a small quantity of sample into beaker.
- Take 10-15 ml into the mouth, hold it for several seconds, and discharge. It is not necessary to swallow the sample to taste it.
- Note the aftertaste, as well as the taste while the sample is in the mouth.

Generally, finished water should have an intensity of 2 or less. A metallic and salty taste may indicate pollution in the water supply.

Turbidity is an optical effect caused by the interception and dispersion of light rays passing through water containing small particles in suspension. It may be caused by silt or clay extracted from soil, surface wash containing suspended organic and mineral matter, precipitated calcium carbonate in hard waters, aluminum hydrate in treated waters, precipitated iron oxide in corrosive water, microscopic organisms, and similar material.

Measurement of turbidity is important because it is one of the visual factors affecting consumer acceptance of water.

- In well water it may indicate the entrance of surface wash, and thus potential contamination.
- In coagulated, filtered water it generally indicates improper operation. Alum floc may be passing the filters or floc may be forming in the clear water well instead of in the coagulation basin.
- In delivered water, turbidity may be due to precipitated calcium carbonate, indicating deposition of scale in pipe lines, or it may be due to iron oxide caused by the corrosion of the pipelines.
- In raw water, it affects the quantity of coagulant required for treatment and shortens filter runs.

Because of the wide variety of materials that cause turbidity in natural waters, it has been necessary to use an arbitrary standard which approximates the conditions found in water. The standard chosen for *Standard Methods* is a suspension of silica in water.

$$1 \text{ mg } SiO_2/1 = \text{one unit of turbidity}$$

The silica used must meet certain specifications as to particle size and the suspensions must be prepared as directed in *Standard Methods*.

The samples collected for color determination may be used for turbidity tests and turbidity may be determined as follows:

- Equipment
  - A commercially available turbid meter may be used in place of the bottle standards.
  - Extra bottles of the same kind as contain the standards.
- Reagents

- Turbidity standards consisting of suspensions of finely divided clay of uniform particle size which, when thoroughly shaken, are equivalent to turbidities of 5, 10, 15, 20, and 25 units may be purchased and stored in bottles similar to the sample bottles.
- Procedure
  - Place a suitable portion of sample into one of the bottles similar to those used for the standards.
  - Shake the sample and standards thoroughly and compare either by transmitted light or by interference to visual perception looking horizontally through the bottles at a ruled or printed paper.
  - If the turbidity is greater than 25 but less than 100, make an appropriate dilution of the sample with distilled water and compare.
  - For turbidities greater than 100, a turbid meter should be used.
  - For very low turbidities see the following low-turbidity test.
  - As the standards are prepared directly in turbidity units, the standard that matches the sample records the turbidity. Readings falling between standards should be interpolated. If the sample has been diluted, the reading must be multiplied by the amount of the dilution factor.

Low-Turbidity Test. The cotton plug filter test is the most sensitive test for determining the amount of coagulated material passing through filter beds and is determined as follows:

- Equipment
  - Analytical balance
  - Muffle furnace or Meker burner
  - Absorbent cotton
  - Silica or platinum dishes
  - Stopwatch
  - Filter tube #46170, 24-27-inch I.D., as listed in Kimble Laboratory Glassware Catalog Sp-75
  - Plastic or copper tubing to connect sample tap to filter tube using a one-hole rubber stopper
  - 50-gallon drum to measure volume of water being filtered
- Procedure
  - Place absorbent cotton inside the filter tube and pack it tightly enough to prevent channeling but not so tightly as to interfere with flow.
  - Open sample tap to an estimated flow of 100 ml per minute.
  - Calibrate flow using a graduated cylinder and a stopwatch. At the suggested flow rate, about 32 hours will be required to fill the 50-gallon drum. Depending on the turbidity, from 3 to 14 days, or from 100 to 500 gallons may be required to produce a weighable sample.
  - Close sample tap, remove the wet cotton, place it in the tared silica dish, dry, and ignite.

- Weigh the residue on the analytical balance. Ignited cotton blanks usually do not give a weighable sample.
- Calculation

$$\text{mg per liter} = \frac{\text{mg of material retained on filter}}{(3.78) \times (\text{number of gallons filtered})}$$

The filterability index has been defined as the ease with which a water can be passed through a given filter. A distinct advantage of this test is that after it has been completed, the filter may be examined for carbon and algae. Once standards have been established, the test is excellent for the control of filtration, and is as follows:

- Equipment
  - Membrane filters
  - Erlenmeyer flasks, 500 ml
  - Graduated cylinder, 500 ml
  - Pyrex filter holder
  - Needle valve
  - Vacuum trap
  - Vacuum pump
  - Stopwatch
  - Manometer or vacuum gauge
- Procedure
  - Collect sample to be tested in a 500-ml Erlenmeyer flask and warm to 20-25°C.
  - Place a membrane filter into the holder, lock into position. Using rubber stopper and tubing, connect the holder to graduate, used as a receiving flask, and to the manometer and vacuum.
  - With the vacuum on and needle valve open, invert the water sample into the funnel with the flask opening held below the water surface. Clamp into position.
  - Adjust the needle valve to bring the vacuum to the desired value, about 7.5 inches of mercury.
  - Measure the time required to filter the sample.
- Calculation

Results may be expressed as the number of milliliters filtered in 5 minutes or some other convenient time. Conversely, the time required to filter a certain fixed volume may also be used.

The U.S. Public Health Service has placed a limit of 5 units of turbidity as the maximum allowable amount in public water supplies, as any greater concentration is readily noted by consumers and indicates unsatisfactory conditions. Coagulation and filtration should always reduce the turbidity to less than 5 units, and in well-operated plants to less than 1.0 unit.

ALKALINITY

Alkalinity is a measure of the alkaline constituents of water. Although bicarbonates represent the major form of alkalinity in natural waters, it may also be present as the carbonate, usually as salts of calcium, magnesium, sodium, and potassium. Alkalinity has major importance in connection with coagulation and corrosion control. Alum is an acid salt which, when added in small quantities to natural water, reacts with the alkalinity present to form floc. If insufficient alkalinity is present to react with all the alum, coagulation will be incomplete and soluble alum will be left in the water. It might, therefore, be necessary to add alkalinity in the form of soda ash or lime to complete the coagulation or to maintain sufficient alkalinity to prevent the coagulated water from being corrosive. Alkalinity can exist as hydroxide, carbonate, or bicarbonate but as this discussion is primarily concerned with the control of coagulation only total alkalinity will be considered.

Alkalinity has little sanitary significance. The U.S. Public Health Service has, however, established standards on chemically treated waters because highly alkaline waters are usually unpalatable, and consumers tend to seek other supplies.

Samples of raw water before coagulation and samples of filtered water are required. The method of determination is as follows:

- Equipment
  - Burette, 50 ml
  - Burette stand
  - 2 Erlenmeyer flasks, 250-ml
  - Graduated cylinder, 100-ml
- Reagents
  - N/50 sulfuric acid
  - Methyl orange or methyl purple indicator solution, 0.5 gm/liter, or bromcresol green--methyl red indicator solution.
- Procedure
  - Fill burette to mark with N/50 sulfuric acid.
  - Using graduated cylinder, measure two 100-ml portions of sample and transfer each to a 250-ml Erlenmeyer flask.
  - To each of the samples in the titration flasks, add two drops of indicator solution.
  - Slowly add from burette N/50 acid to one of the test portions in the titration flask, mixing thoroughly by rotating the flask.
  - Continuously compare the colors in the two flasks as the acid is added, and at the first appearance of a difference in color between the two flasks, stop the addition of N/50 sulfuric acid.

- Read the burette
- Calculation

$$\text{mg/l total alkalinity as } CaCO_3 = \frac{(\text{ml std acid}) (1000)}{\text{ml of sample}}$$

The determination of alkalinity provides an estimate of the alkaline constituents in water. If all the alkaline constituents are present as salts of calcium and magnesium, the alkalinity will equal the hardness. If the alkalinity is greater than the hardness it must mean there are alkaline salts of metals other than calcium and magnesium present, generally, sodium or potassium salts. If the alkalinity is less than the hardness, there must be salts of calcium or magnesium present that are not carbonates. These are usually sulfates. An alkalinity of less than 100 mg/l is desirable for water used for domestic purposes.

Coagulation generally requires a concentration of alkalinity equal to half the amount of alum added to produce good floc. Thus, two grains per gallon of alum (34.2 mg/l) requires 17.1 mg/l of alkalinity to form foc. As coagulation destroys the alkalinity, the water becomes more corrosive and unless an excess of alkalinity was present before coagulation, soda ash or lime must be added to the filtered water to prevent corrosion. Generally, if the concentration of alkalinity before coagulation is equal to or greater than the alum dose, no increase in the corrosive quality of the water will be caused by coagulation.

The relation between alkalinity and pH is a determining factor in whether or not a water will form scale in the distribution system.

Coagulation

To insure optimum coagulation of water, close control of the amounts of chemicals used is essential. A slight excess or an insufficient dosage or any one or all of these chemicals will result in unsatisfactory coagulation, and thus incomplete and ineffective treatment. Since the character of the raw water may change appreciably from day to day, it is desirable that the correct dosages of chemicals required be determined daily. Raw water should be collected in clean glass bottles at a point representative of the water to be treated in plant practice. The method of determination is as follows:

- Equipment
  - Stirring device, to operate at rate of 10-15 Rpm
  - Glass jars or beakers, for us with stirring device
  - Filter funnels
  - Filter paper
  - Equipment for the determination of color, turbidity, pH, and alkalinity
- Reagents—concentrations should be 1,000 mg/l
  - Coagulant

- Other chemicals used in plant coagulation
- Procedure. It is desirable to simulate in the laboratory tests the conditions of mixing, sedimentation time, and so on that actually exist in the plant.
  - Measure known volumes of raw water, usually 500 ml, into each of six jars or beakers and place jars in the stirring device.
  - Starting with first jar on the left, add gradually increasing doses of chemicals used for coagulation. Select the series of doses so that the first jar will represent undertreatment and the last jar will represent overtreatment.
  - Start the stirring device and stir sample at a rate of 10-15 rpm.
  - Stir for 15 minutes.
  - Observe floc formation during the stirring period. Record those test portions showing good to excellent floc formation.
  - Allow floc to settle, usually 15 to 60 minutes.
  - Withdraw portions of clear, settled water from each test portion.
  - Determine color, turbidity, pH, and alkalinity of each portion.

The test portion giving maximum reductions of color and turbidity and the most satisfactory floc formation will indicate the dosages of chemicals required for optimum treatment of the raw water. The determination of alkalinity and the pH value will indicate the corresponding values which should be found in the sedimentation basin effluent following effective treatment.

Waters having an alkalinity of less than 100 mg/l coagulate best usually at pH values between 5.5 and 7.0. Waters of greater than 100 mg/l alkalinity usually give optimum coagulation results at pH values between 7.0 and 7.6.

FLUORIDE

Fluorine is one of the most active elements known. It forms simple fluoride compounds as well as many complex compounds in combination with other ions. Some of these dissociate in water to yield the fluoride ion. Compounds of fluoride generally used for addition to public water supplies include sodium fluoride, sodium silicofluoride, hydrofluosilicic acid, calcium fluoride, and ammonium silicofluoride.

A concentration of from 1.0 to 1.2 mg/l of fluoride ion in drinking water helps prevent dental caries, especially in the teeth of children. Fluoride ions, if present in water in concentrations greater than about 2.0 mg/l, may cause mottling of the tooth enamel of consumers. Samples of raw and finished water from the distribution system are required.

The standard acceptable procedures for the measurement of fluorides in water are the calorimetric methods based on the bleaching of a preformed color by the fluoride ion. Zirconium ion and alizarin dye unite to form a reddish colored "lake" dye which is reduced to a yellowish color if the amount of zirconium is decreased. Fluoride ion will combine with zirconium ions to form a colorless complex anion

($ZrF_6^{=}$), and the intensity of the color of the lake decreases accordingly. Interfering ions must be removed prior to the test. A commercial comparator can be used to evaluate the resulting test colors. The method is as follows:

- Equipment
    - Commercial test kit (a Hach colorimeter or a Hellige Aqua Tester is acceptable)
    - Erlenmeyer flasks, 126 ml
    - Mohr pipettes, 5 ml, with 0.1-ml graduations and 10 ml with 0.1-ml graduations
    - Beakers, 600 ml
    - Graduated cylinders, 50 ml
    - Volumetric flasks, 100 ml
    - Interval timer
    - Spot plate
    - Funnels, short stem, 2.5 inch diameter
    - Filter paper, Whatman #40, 11-cm diameter
- Reagents
    - Acid zirconium-alizarin reagent
    - Orthotolidine solution
    - Sodium arsenite solution, 0.5 percent
    - Acid-dichromate cleaning solution
    - Activated carbon
    - Sodium arsenite solution
- Interfering substances

   In any method for fluorides which involves the decolorization of a lake, the presence of residual chlorine and phosphate should receive consideration. Residual chlorine tends to bleach the alizarin color but may be eliminated by the addition of a slight excess of sodium arsenite solution. Phosphates, especially metaphosphates (such as Calgon), simulate the presence of fluoride and tend to give higher readings. Excessive amounts of color or turbidity in the water produce off-shades with the fluoride reagent.
- Pretreatment of sample
    - Test the sample for the presence of residual chlorine by placing a few drops of the sample in a spot plate and adding one drop of reagent. If a yellow color develops indicating the presence of residual chlorine, place 100 ml of sample into a 125-ml Erlenmeyer flask, add two to three drops of sodium arsenite solution and mix thoroughly. Recheck for the presence of residual chlorine, and if still present, repeat the preceding procedure until the addition of reagent shows no color.
    - If the presence of Calgon is suspected, boiling the sample for 10 minutes will eliminate this interference which is due to metaphosphate. The sample must be cooled and the original volume restored by the addition of distilled water before proceeding with the fluoride test.

- In the presence of excessive amounts of color or turbidity which may be difficult to compensate for, even by the use of a sample blank, activated carbon may be used to remove the color. Filtration through Whatman #40 filter paper should remove the turbidity. Do not use chemical coagulants. Check any treatment to insure that the fluoride content of the water is not affected.
- Procedure. The following is applicable for the Hellige aqua Tester.
  - Rinse Nessler tubes with acid-cleaning solution followed several times by distilled water.
  - Measure 50 ml of the pretreated sample into a 125-ml Erlenmeyer flask. Adjust the temperature to 23° to 27°C.
  - Add exactly 2.5 ml of the acid zirconium-alizarin reagent, mix well, and set times for exactly one hour.
  - At the end of the hour, pour the treated sample into the measuring tube and read this in the comparator against a second tube filled with untreated sample and placed under the colored glass standards.
  - Obtain readings by revolving the color disc until a color match is obtained between the tubes.
  - Values higher than those marked on the disc may be determined by diluting the sample with distilled water before the fluoride reagent is added.
  - Specific procedures listed by the manufacturer should be followed if a different comparator is used.
- Calculation
  - mg/fluoride = $\frac{[(\text{Disc reading (50)}]}{\text{ml of sample}}$
  - Report readings to the nearest 0.05 mg/l.

The fluoride ion determined represents the total concentration including fluoride naturally present in the raw water plus that added by fluoridation. The raw water should be analyzed separately to determine the amount of natural fluorides present so that proper amounts of supplemental fluorides can be added.

Where fluoridation is practiced, it is necessary to maintain surveillance on the finished water to insure that proper amounts of chemicals are being fed. The usual practice is to collect samples on the distribution system as well as at the treatment plant.

HARDNESS

Water is a strong and universal solvent and dissolves varying amounts of different mineral substances. These do not affect the sanitary quality but are of importance in the domestic use of water, particularly for laundry and boiler purposes. Calcium and magnesium salts, the principal mineral constituents by which waters are characterized as *hard*, combine with soap and precipitate as insoluble soap curds. Until all the calcium and magnesium are precipitated, no appreciable lather or washing action is obtained. Calcium and magnesium are generally present in water as soluble bicarbonate salts. When the water is heated the less soluble carbonate salts of these metals are precipitated because carbon dioxide is driven off by heat. This is the source of scale in distributing systems and hot water heaters.

Hard waters are as satisfactory for human consumption as soft waters. They are not satisfactory for use with soap cleansers, however, because of the formation of soap curd. Synthetic detergents usually are unaffected by the hardness of the water. Hardness is also an important factor in the formation of scale in boilers, water heaters, and distribution systems.

Hardness is normally determined in finished waters collected from the plant or from the distribution system. The EDTA titration method is now universally used:

- Equipment
    - Porcelain casserole, 100 ml
    - Glass stirring rod
    - Graduated cylinder, 50 ml
    - Assorted pipettes
    - Burette, 50 ml
- Reagents
    - Buffer solution—Dissolve 16.9 gm ammonium chloride in 143-ml conc, ammonium hydroxide; add 1.24 gm of magnesium salt of EDTA and dilute to 250 ml with distilled water
    - Eriochrome black T indicator
    - Standard EDTA solution: 1.0 ml = 1.0 mg $CaCO_3$
- Procedure
    - Fill burette to mark with EDTA solution.
    - Using a graduated cylinder, measure 50 ml of sample into the porcelain casserole.
    - Add 1 to 2 ml of buffer solution.
    - Add one to two drops of indicator solution.
    - Add EDTA solution slowly, stirring constantly with the glass rod, until the last reddish tinge disappears and the sample becomes bluish.
- Calculation

$$\text{mg/l hardness as } CaCO_3 = \frac{(\text{ml EDTA}) (1000)}{\text{ml of sample}}$$

Water most satisfactory for domestic use contains about 50 milligrams per liter of hardness. Water with a hardness of 300 or greater is not suitable for ordinary use. Very soft waters, having a hardness less than 30 mg/l, are likely to be very corrosive. Such waters are generally treated with lime which increases the hardness.

Iron and Manganese

Both iron and manganese create serious problems in public water supplies. Under reducing conditions they are relatively soluble in natural water, but upon exposure to air they precipitate out of solution.

There is no indication at the present time of any harmful effects to humans from drinking waters containing iron and manganese. Waters containing these elements, however, on exposure to air become turbid and colored, making them aesthetically unacceptable. Since they precipitate out of solution, they interfere with laundering operations and often deposit on plumbing fixtures.

The U.S. Public Health Service Standards recommend that in public water supplies, the concentrations of iron and manganese should be limited to 0.3 mg/l and 0.05 mg/l, respectively. Samples of raw and finished water are required for analysis. The method for determination is the 1.10-Phenanthroline method for iron.

- Equipment
  - Graduated cylinder, 50 ml
  - Erlenmeyer flask, 125 ml
  - Pipettes, assorted
  - Glass beads
  - Bunsen burner
  - Nessler tubes, short form, 100 ml
  - Light comparator stand
- Reagents
  - Hydrochloric acid, conc.
  - Hydroxylamine reagent. Dissolve 10 gm in 100 ml distilled water.
  - Ammonium acetate buffer solution. Dissolve 250 gm in 150 ml distilled water. Add 700 ml glacial acetic acid and dilute to 1 liter.
  - Phenanthroline solution. Dissolve 0.1 gm in 100 ml distilled water. Heat to 80°C to aid solution but do not boil. Discard if it darkens on heating.
  - Two standard iron solutions. 1.0 ml = 0.01 mg Fe and 1.0 ml = 0.001 mg Fe
- Procedure
  - Wash all glassware with conc. HCl and rinse several times with distilled water.
  - Using the graduated cylinder, measure 50 ml of sample into a 125-ml Erlenmeyer flask.
  - Add 2 ml conc. HCl, 1 ml hydroxylamine hydrochloride, and a few glass beads.
  - Using the Bunsen burner, heat to boiling and continue boiling until volume is reduced to 15 to 20 ml.
  - Cool to room temperature and transfer to a 100-ml short-form Nessler tube.

- Add 10 ml acetate buffer solution, 2 ml phenanthroline solution, and dilute to mark with distilled water.
- Mix thoroughly and allow 10 to 15 minutes for color development.
- Compare against a series of standards which have been carried through the same procedure. Compare by looking down vertically through the tube held against an illuminated light comparator. Color standards are stable for periods up to six months if protected from evaporation.

- Calculation

$$\text{mg/l Fe} = \frac{(\text{mg Fe})(1,000)}{\text{ml of sample}}$$

Periodate Method for Manganese

- Equipment
    - Pipettes, assorted
    - Graduated cylinder, 250 ml
    - Erlenmeyer flask, 500 ml
    - Glass beads
    - Bunsen burner
    - Nessler tubes, tall form, 50 ml
    - Light comparator stand
- Reagents
    - Sulfuric acid, conc
    - Nitric acid, conc
    - Phosphoric acid, syrupy, 85 percent soln
    - Potassium metaperiodate, KIO_4
    - Silver nitrate, $AgNO_3$
    - Manganous sulfate standard solution, 1.0 ml = 0.05 mg Mn
- Procedure
    - Using graduated cylinder, measure 200 ml of sample into a 500-ml Erlenmeyer flask.
    - Add 5 ml conc. H_2SO_4, 5 ml conc. HNO_3, and a few glass beads.
    - Heat over Bunsen burner until dense white fumes of SO_3 appear in the flask.
    - Cool, carefully add 40 ml distilled water and cool again.
    - Add 5 ml HNO_3, 5 ml H_3PO, and mix.
    - Add 0.3 gm KIO_4 and 20 mg $AgNO_3$.
    - Heat to boiling and keep just below boiling point for 10 minutes.
    - Cool and dilute to 50 ml in tall-form Nessler tube.
    - Compare visually using standards prepared by treating the various amounts of standard Mn solution in the same way. Color standards are stable for periods up to six months if protected from dust and evaporation.
- Calculation

$$\text{mg/l Mn} = \frac{(\text{mg Mn})\,(1{,}000)}{\text{ml of sample}}$$

Iron and mangese concentrations are important considerations when new water supplies are being developed. A potential supply may be rejected when these concentrations are excessive. Results of these analyses are used to formulate the type and extent of treatment necessary to bring iron and manganese levels to within recommended limits. Results can be used to detect the presence of iron-fixing bacteria and corrosion in distribution systems.

Marble Test

To reduce the corrosive action of water containing dissolved carbon dioxide, it is necessary to increase the pH value by the addition of an alkali. If lime is added, a pH value may be obtained at which corrosion will be reduced to a minimum and a protective coating of calcium carbonate be precipitated on the mains and pipes. If the correct pH value and the corresponding concentration of alkalinity are maintained, a condition of stability should result without further deposition of calcium carbonate, or solution of the coating previously precipitated. The purpose of the marble test is to determine in the laboratory the correct pH value and alkalinity required by a particular water to prevent corrosion and permit formation of a protective coating of calcium carbonate.

The same considerations described under the sanitary significance of alkalinity hold true. Samples should be representative of the supply to be treated and should be collected in glass-stoppered bottles of Pyrex or similarly resistant glass of 250-ml capacity. The sample bottle should be filled so that no entrained air will collect under the dropper. The method of determination is:

- Equipment
    - Glass-stoppered BOD bottles—300-ml capacity of Pyrex or other resistant glass
    - Whatman #50 filter paper, 18.5-cm diameter
    - Glass funnel, 125 mm diameter
    - Glass siphon
    - Equipment for determination of alkalinity and pH
- Reagents
    - Precipitated calcium carbonate, reagent grade
    - Reagents for determination of alkalinity and pH
- Procedure
    - Fill the bottle with the sample, without agitation.
    - Add about 1 gram of calcium carbonate, and replace the stopper without any entrapment of air.
    - Mix by shaking at frequent intervals for at least three hours to aid in the solution of the calcium carbonate and the attainment of equilibrium.

- Allow the sample to settle overnight and carefully siphon about half of the clarified supernatant into a flask. It is important that no agitation of the sample occurs.
- Filter the siphoned supernatant through Whatman #50 filter paper, discarding the first portion, and determining total alkalinity on 100 ml of the remainder.
- Determine pH value of the unfiltered supernatant.
- Calculation. The results are reported as the pH value and the alkalinity at calcium carbonate equilibrium.

The increase in the total alkalinity and in the pH value indicates the corresponding values which should be obtained by the addition of lime in the treatment of water to eliminate corrosive action and to provide for the deposition of a protective coating. In the treatment of the supply, the dosage of lime should be controlled to maintain the total alkalinity and the pH value at the concentration determined by these tests.

Mud Balls

An aggregation of solids in sand filter beds termed *mud balls* generally indicates incomplete washing due to too short a wash period or an insufficient flow of wash water. These aggregates can be prevented by the agitation of the sand beds during the wash cycle, either mechanically or by water pressure. Once present, these mud balls can be removed by screening the sand surface during the wash period or by scraping the sand bed surface with shovels. If not removed, small mud balls may agglomerate into larger balls and work their way down through the sand, decreasing the efficiency of a filter.

There is no sanitary significance associated with the presence of mud balls in a sand filter. Samples of washed filter sand are collected as described here and examined.

- Equipment
  - #10 mesh (0.1-inch) wire sieve
  - Sheet metal sampling tube, 3 inches in diameter, 6 inches in length, with a wooden base which can be closed by an attached handle.
  - Graduated cylinder, 250 ml
  - Vessel to contain water for washing wire sieve
- Procedure
  - Wash sand bed in usual manner.
  - Lower the water level to at least 12 inches below the sand surface.
  - Collect at least four samples of sand from different portions of the sand bed by inserting the sampling tube vertically its full length into the sand. Move the handle to close the wooden base and transfer the tube to a portable container.

- Place portions of the collected sand in the sieve and then raise and lower it gently in a vessel of water so that the sand grains are washed from the mud balls which are retained on the sieve. Repeat the washing process until all the mud balls have been separated from the sand.
- Determine the volume of mud balls by transferring them carefully to a graduated cylinder partly filled with water and noting the apparent increase in the volume of the water in the graduate due to addition of the mud balls.
- Calculation

$$\% \text{ Mud balls} = \frac{\text{Volume of mud balls}}{\text{Volume of sand examined}} \times 100$$

where the volume of sand examined is a constant determined by the size of the sampling container. Using the container described previously for four samples, the constant in cubic centimeters would be 2,760.

Most mud balls remain near the sand surface. If only a small percentage is found in the top 6 inches, the volume of mud balls will probably be low throughout the entire depth of the sand bed. The condition of filter beds tested by this method of Baylis can be classified according to Table 4-4.

Table 4-4 Interpretation of Test for Mud Balls

Percent by volume of mud balls	Condition of filter bed
0.0 to 0.1	Excellent
0.1 to 0.2	Very good
0.2 to 0.5	Good
0.5 to 1.0	Fair
1.0 to 2.5	Fairly bad
2.5 to 5.0	Bad
5.0	Very bad

RESIDUAL CHLORINE

The purpose of chlorinating public water supplies is to prevent the spread of waterborne diseases. Chlorine, however, when added to water, reacts with organic and other substances including bacteria. It is necessary to add sufficient chlorine to react with all the various substances present and still leave an excess or residual to insure destruction of all of the bacteria. Residual chlorine may be present in the free or available state, which has a very rapid disinfecting power. It may be combined with ammonia to form less active chloramine; or it may be absorbed by organic matter to form relatively inactive chlororganic compounds which have little or no disinfecting power. In the chlorination of water supplies, it is important that the residual chlorine be present as free available chlorine rather than as the less active combined chlorine.

The presence of free available chlorine in a water at least 10 minutes after chlorine has been introduced and thoroughly mixed insures destruction of all harmful bacteria. Samples must be collected in clean bottles and if a test kit is used for color comparison, the cells must be clean. Any dirt on the glassware will react with the chlorine in the sample and give low results. If possible, samples should be collected at a point where the chlorine has been in contact with the water for at least 10 minutes. If this cannot be done, the sample should be allowed to stand for a sufficient time to make the total contact period 10 minutes.

The OTA test will differentiate and measure both free and combined chlorine residuals. The method is based upon the fact that free chlorine residuals react instantaneously with orthotolidine to produce a yellow compound, whereas combined chlorine reacts much more slowly, requiring 5 minutes at 70°F for full color development. Also, the addition of a reducing agent (sodium arsenite) will neutralize any residual chlorine either free or combined without affecting any color initially formed. Free chlorine residuals can be determined by adding sodium arsenite solution within 5 seconds after the orthotolidine is added. Time is allowed, thereby, for the free chlorine to oxidize the orthotolidine, but the slower-acting combined chlorine is eliminated before it has a chance to react. Comparison of this color with permanent standards will give a quantitative value for the free residual present.

Orthotolidine reagent when added to another portion of the sample and allowed to react for 5 minutes, will develop color due to both the free available and combined chorine residuals, which can be compared with permanent standards.

Since permanent standards are prepared to simulate the color due to chlorine in water which is free from color, turbidity, and interfering substances, a correction for these constituents in the sample must be made when making a color comparison. This is done by using a blank sample (cell B) to which sodium arsenite is added first, thus destroying all residual chlorine, followed by the addition of orthotolidine reagent, which may or may not develop color due to interfering substances.

The significance of these three cells can be summarized as follows:

    Cell A = free chlorine and interferences
    Cell B = interferences
    Cell OT = free chlorine, combined chlorine and interferences

Using the readings from these three portions, it is possible to calculate values for the free available and combined chlorine residuals.

- Equipment
    - Chlorine test kit with permanent standards (Hellige, Taylor, Wallace and Tiernan or others are satisfactory)

- Extra cells for test kit
- Interval timer
- Reagents
  - Orthotolidine (OT) reagent
  - Arsenite reagent
- Procedure. The following applies to Hellige and Wallace and Tiernan comparators.
  - Hold samples for the required 10-minute contact period unless a longer time has elapsed between the application of chlorine and the collection of sample.
  - Label three kit cells as A, B, and OT.
  - To cell OT add one dropperful of OT reagent and fill to mark with sample. Use a timer and set aside for 5 minutes. If below 70°F (20°C), warm the sample and OT in the cell to 70°F in a bath of a slightly warmer water.
  - To cell A add one dropperful of OT reagent, fill to mark with sample and immediately (within 5 seconds) add dropperful of arsenite reagent. Mix.
  - To cell B add one dropperful of arsenite reagent, fill to mark with sample, and add one dropperful of OT reagent. Mix.
  - For the free available chlorine residual, place cell A in left-hand slot of kit and cell B in right-hand slot. (In a Hellige comparator, place cell A on the right and cell B on the left. In general, cell B is placed in line with the permanent color standards.) Rotate disc until color matches on both sides of field when viewed through eyepiece.
  - For the combined available chlorine residual, at the end of 5 minutes, place cell OT in the left-hand slot of kit and cell A in the right-hand slot of the kit (in a Hellige, cell C)T is placed on the right and cell on the left), and rotate disc until color matches on both sides of field as viewed through eyepiece.
- Calculation. Results are read directly from the disc and reported as mg/l free available chlorine residual or combined available chlorine residual.

The disinfecting power of chlorine depends upon the form of residual chlorine present, the contact time, the temperature, and the pH value of the water. For example, if the pH value is less than 8.0, then 0.2 mg/l of free residual chlorine will destroy bacteria in a 10-minute contact period at all temperatures. To accomplish the same results with combined chlorine, a residual of 1.0 at pH 6.0, 1.5 at pH 7.0, and 1.8 at pH 8.0 must be maintained with a contact period of 60 minutes, and this residual must be varied as the temperature of the treated water varies.

HYDROGEN-ION CONCENTRATION—pH

pH expresses the intensity of the acid or alkaline reaction of a solution in terms of the hydrogen-ion concentration but is not a measure of the total concentration of acid or alkali present. It is the logarithm of the reciprocal of the hydrogen ion concentration in moles per liter. Water containing no acid or alkali has a pH value of 7.0, which is termed the *neutral pH value*. Addition of strong acids, such as sulfuric or hydrochloric, markedly reduce the pH value while an equal amount of weak acid such as carbonic acid only slightly lowers the pH value. Similarly, alkali increases the pH value to above 7.0 and the degree of change depends on the intensity and amount of alkali added. Thus, pH value below 7.0 indicate acidity, pH value 7.0 indicates neutrality, and pH values above 7.0 indicate alkalinity. Most natural waters have pH values between 5.5 and 8.6.

In the field of water treatment, pH is a factor which must be considered in chemical coagulation, disinfection, water softening and corrosion control.

Samples for pH determination must be collected carefully. Very small quantities of contaminants can affect the results markedly. Hands that have handled lime or alum can carry enough chemical to affect the results if they come in contact with the sample. Samples must be collected without agitation since loss of carbon dioxide gas can change the pH value. (See Figure 4-1).

pH can be measured either colormetrically or electrometrically. The colormetric method described requires less expensive equipment but is subject to interferences which at times render it useless. In these circumstances an electro pH meter must be used.

SOLUTION		pH VALUE	CONCENTRATION IN GRAM PER LITER	
			HYDROGEN ION	HYDROXYL ION
5% Sulfuric acid →	INCREASING ACIDITY	0	$1 = 1 \times 10^{0}$	$0.00000000000001 = 1 \times 10^{-14}$
		1	$0.1 = 1 \times 10^{-1}$	$0.0000000000001 = 1 \times 10^{-13}$
		2	$0.01 = 1 \times 10^{-2}$	$0.000000000001 = 1 \times 10^{-12}$
Lemon juice →		3	$0.001 = 1 \times 10^{-3}$	$0.00000000001 = 1 \times 10^{-11}$
Orange juice →		4	$0.0001 = 1 \times 10^{-4}$	$0.0000000001 = 1 \times 10^{-10}$
		5	$0.00001 = 1 \times 10^{-5}$	$0.000000001 = 1 \times 10^{-9}$
		6	$0.000001 = 1 \times 10^{-6}$	$0.00000001 = 1 \times 10^{-8}$
Milk → Water → Body fluids →	NEUTRAL	7	$0.0000001 = 1 \times 10^{-7}$	$0.0000001 = 1 \times 10^{-7}$
Egg whites →	INCREASING ALKALINITY	8	$0.00000001 = 1 \times 10^{-8}$	$0.000001 = 1 \times 10^{-6}$
		9	$0.000000001 = 1 \times 10^{-9}$	$0.00001 = 1 \times 10^{-5}$
Milk of magnesia →		10	$0.0000000001 = 1 \times 10^{-10}$	$0.0001 = 1 \times 10^{-4}$
		11	$0.00000000001 = 1 \times 10^{-11}$	$0.001 = 1 \times 10^{-3}$
Milk of lime →		12	$0.000000000001 = 1 \times 10^{-12}$	$0.01 = 1 \times 10^{-2}$
		13	$0.0000000000001 = 1 \times 10^{-13}$	$0.1 = 1 \times 10^{-1}$
4% Caustic soda →		14	$0.00000000000001 = 1 \times 10^{-14}$	$1 = 1 \times 10^{0}$

Figure 4-1 The pH scale

- Equipment
  - pH comparator
  - Standard color discs for comparator, covering pH ranges anticipated
  - Two glass tubes to fit comparator
  - Dropper pipettes
- Reagents
  - Indicator solutions, covering pH ranges anticipated
- Procedure—for comparator kits employing colored glass discs
  - Fill two comparator tubes to the mark with the sample to be examined.
  - To one of the tubes, add the required amount of indicator, exactly as specified by the manufacturer.
  - Place the other tube containing the untreated sample in the slot back of the colored glass discs. Place the tube containing the indicator in the other slot.
  - Look through the eyepiece while holding the apparatus up toward the blue sky (not sunlight) or daylight fluorescent lamp and rotate the disc until the colors match, as seen through the eyepiece.
- Interferences—Color and turbidity are the chief interferences with this method. If these together or singly are not great, the interference may not be sufficiently serious to prevent a correct reading. The organic materials in water which produce color usually adsorb the color indicator dyes. The result is an off-shade which is difficult to match with standards. Turbidity due to organic substances would interfere in the same manner; mineral turbidity usually does not adsorb indicator dyes and thus presents no great problem unless it is very intense.
- Calculation. Read the pH value from the exposed number on the disc. pH values less than 7.0 indicate an acid reaction, the intensity increasing ten times for each drop of one unit. A pH value of less than 6.0 indicates a definitely aggressive or corrosive action of the water toward metals used in service pipes. Charts are available for use in conjunction with pH values and alkalinity to determine whether scale will form on distributing pipes.

For effective coagulation of different waters by alum, various pH values will be found as optimum and can be used for control of this process.

BACTERIA COLIFORM GROUP

Multiple-Tube Fermentation Techniques

All members of the coliform group ferment lactose (milk sugar) with the formation of acid and gas, grow aerobically (in the presence of oxygen), and do not form spores. These cultural characteristics are the basis for the routine test for the presence of this group in a water sample. No attempt is made to differentiate between the numerous species comprising the group since the presence of any member of the group is considered by *Standard Methods* to be indicative of the presence of sewage pollution of the water. The test gives an estimate of the number of bacteria of the coliform group present in a given volume of water as an index of the degree of pollution. This value is generally referred to as the *most probable number* (MPN).

The U.S. Public Health Service has established standards relative to the frequency and number of samples which should be examined as well as to the quality of the water as revealed by the results of the examination.

- Samples must be collected in sterile plastic containers, or sterile, glass-stopped, glass bottles, following all directions as given in the foregoing section on sampling in order to avoid the contamination of the sample.
- Sampling points must be at locations which will provide samples representative of the water supply.
- Avoid dead ends and leaky taps.
- Allow water to run from the tap for two or three minutes before collecting samples.

The method of determination is:

- Equipment
  - Harvard trip balance
  - Pressure cooker autoclave to operate at 15 lbs pressure, 121°C
  - Hot-air sterilizing oven to operate at 170°C
  - Incubator equipped to maintain constant temperature between 34°C and 36°C in incubation chamber
  - Graduated cylinder, 500 ml
  - Baker, 1,000 ml
  - Glass stirring rod
  - Bunsen burner
  - Cotton plugs or plastic closures for fermentation tubes
  - Milk-diluting pipette, graduated to deliver 1.0 ml and 1.1 ml
  - Pipettes, volumetric, transfer 10 ml
  - Fermentation tubes for 10-ml volumes of sample—culture tubes without lip, approximately 175 × 22 mm or screw-top tubes of same size
  - Fermentation tubes for 1-ml volumes of sample—culture tubes without lip, approximately 150 × 18 mm or screw-top tubes of same size
  - Inner tubes—culture tubes without lip, approximately 50-75 mm long and 5-10 mm diameter
  - Baskets or racks for holding media
  - Inoculating loop for transferring cultures—24-gauge wire loop not less than 3 mm in diameter
- Reagents
  - Dehydrated lactose broth, for water examination
  - Dehydrated brilliant green lactose bile broth 2 percent, for water examination

- Preparation of media
    - Lactose broth. The medium after inoculation with the sample must contain 0.5 percent each of lactose and peptone. Thus, in tubes to which 10 ml of water are to be added, the medium must be made up to double strength.
        1. Weigh 13 gm of dehydrated lactose broth.
        2. Measure 500 ml of distilled water in a graduated cylinder.
        3. Heat 400 ml of the distilled water to boiling in a beaker.
        4. Suspend the weighted lactose broth in the remaining 100 ml of cold distilled water.
        5. With constant stirring, add the suspension to the boiling water and dissolve completely by stirring vigorously.
        6. To each of the large tubes (175 × 22 mm), insert one of the small tubes in an inverted position.
        7. Add 10 ml of the medium to each tube.
        8. Close each tube with a cotton plug or loosely with a screw cap.
        9. Place tubes in baskets and put them in autoclave.
        10. Sterilize for 15 minutes after the pressure has reached 15 lbs/sq inch and the temperature 121°C. Shut off heat at the end of the sterilization period.
        11. Remove from the autoclave as soon as the pressure has returned to zero to prevent decomposition of the sugars by prolonged heating. Cool the tubes and tighten the screw caps. The total time for heating, sterilizing, and cooling should not exceed 40 minutes.
        12. In tubes to which 1-ml of the sample is to be added, normal-strength medium is used. For this purpose in step (1), weigh 6.5 gm, and in step (6) use smaller tubes (150 × 18 mm).
    - Brilliant green lactose bile broth. The procedure is the same as for the preparation of lactose broth except that 20 gm dehydrated medium are used for 500 ml of distilled water. Only single-strength medium is prepared as no direct inoculations are made. Only the 150 × 18 mm tubes with inner tubes are used.

Unless the medium prepared is to be used within a week, it is better to use plastic screw cap fermentation tubes to avoid evaporation of water during storage. Storage of cotton plug tubes in a refrigerator is not recommended because the medium absorbs air during storage and the air is released during incubation. Bubbles of air formed in this manner are often falsely assumed to be gas formed by fermentation of lactose.

- Sterilization of glassware
    - Wrap pipettes in Kraft paper or place in metal pipette can.
    - Sterilize in oven by heating at 170°C for 1 hour.
- Procedure
    - Presumptive test

1. Shake sample violently in an up and down motion 25 times.
2. In an aseptic manner, inoculate each of five large fermentation tubes containing double strength lactose broth with 10 ml of sample.
3. Inoculate one small fermentation tube containing lactose broth with 1 ml of sample.
4. Inoculate one small fermentation tube containing lactose broth with 1/10 ml of sample.
5. Place all fermentation tubes in incubator maintained at $35°C \pm 0.5°C$.
6. At end of 24 hours observe if gas has formed in the inner tube of each of the fermentation tubes.
7. Perform confirmatory tests on all tubes in which gas has formed and replace rest of tubes in incubator.
8. At end of 48 hours observe if gas has formed in the inner tube of each of the remaining lactose tubes.
9. Perform confirmatory tests on all tubes in which gas has formed.

- Confirmatory test. The production of gas in the lactose broth does not necessarily indicate the presence of bacteria of the coliform group because there may be other bacteria present which ferment lactose. If cultures from those lactose broth tubes high show gas is transferred to brilliant green bile broth, the bacteria other than coliform organisms are inhibited by the brilliant green bile broth. Any gas which is produced in these tubes can be assumed to indicate the presence of organics of the coliform group.
  1. Select each fermentation tube showing gas at 24 or 48 hours (steps 6 and 8 of presumptive test) and transfer a loopful of broth to a fermentation tube containing brilliant green bile broth.
  2. Place in incubator for 24 hours.
  3. Examine for presence of gas. If gas is present, the tube may be recorded as positive and discarded; if no gas is present it should be reincubated for another 24 hours and reexamined.
  4. If gas is present at the end of the second 24 hours the tube may be considered positive; if no gas is present, it is negative.
- Calculation. Any lactose broth fermentation test showing gas formation after 24 or 48 hours of incubation, confirmed by gas formation in the confirmatory medium after 24 or 48 hours, indicates the presence of bacteria of the coliform group in the corresponding volume of sample examined. Using the number of positive and negative tubes in each dilution, calculate the MPN per 100 ml of sample from the following. The U.S. Public Health Service recommends that the presence of coliform organisms in a potable water should not exceed 10 percent of the standard 10 ml-samples examined in any single month.

Bacteria of the Coliform Group—Membrane Filter Technique

The membrane filter procedure makes possible a more rapid and more reproducible determination of coliform densities using much larger volumes of sample than the multiple-tube fermentation technique. It provides a direct enumeration of the bacterial density rather than a statistical estimate. However, it does have limitations, since turbidity will interfere as will a large number of noncoliform organisms in the water. The method of determination is as follows:

- Equipment
  - Milk-dilution bottles, Pyrex with rubber stoppers
  - Low-power microscope or other optical device giving a 10-15X magnification
  - Petri dishes, glass or presterilized plastic, 60-mm diameter, 15-mm depth
  - Filter-holding assembly
  - Pyrex vacuum flask, 1,000 ml
  - Vacuum source
  - Stainless steel forceps, smooth inner surface, round-tipped
  - Filter membranes, 47-mm diameter, grid marked
  - Adsorbent pads, 47-mm diameter
- Reagents
  - Dehydrated MF—Endo Broth
  - Buffered dilution water
- Preliminary operations
  - Preparation of medium
    Add 20 ml ethyl alcohol, 95 percent, to 1 liter distilled water.
    Dissolve the appropriate amount of dehydrated medium in this solution, as described on the bottle label.
    Heat the medium with continuous stirring until the boiling point is reached. Do not boil!
    Store the medium in a sterile bottle at a temperature of 2-10°C, and discard any unused medium after 96 hours.
  - Sterilization
    Filter holder. Wrap the funnel and base of the filtration unit separately in Kraft paper and autoclave at 15 lbs/sq inch pressure at a temperature of 121°C for 10 minutes.
    Filter membrane. Autoclave grid marked membrane filters and adsorbent pads as received in their Kraft paper envelopes for 10 minutes at 121°C.
    Forceps. Sterilize before use by dipping in 95 percent ethyl alcohol and igniting the fluid.
    Glassware. Sterilize in oven 170°C for 1 hour.
    Buffered dilution water. Autoclave for 20 minutes after pressure has reached 15 lbs at a temperature of 121°C.
- Procedure
  - Place a single sterile absorbent pad in each sterile Petri dish. Add 1.8 ml of prepared Endo medium to each dish, replace covers, and mark with sample identity.

- Insert filter holder base aseptically into the neck of a 1 liter side-arm vacuum flask.
- Using alcohol-flamed sterile forceps, place a sterile HA filter disc, grid side up, on the filter holder base.
- Carefully place the filter holder funnel in place and lock securely.
- Pour water sample of an appropriate size (100-500 ml for finished waters) into funnel and draw it through the membrane filter into the filter flask by vacuum.
- Rinse the funnel three times with 20-30 ml volumes of sterile buffered dilution water.
- Remove the funnel and carefully transfer the filter to the prepared Petri dish with sterile forceps. "Roll" the filter onto the absorbent pad to avoid entrapping air bubbles.
- Incubate the dishes in an inverted position for 18 to 22 hours at 35°C + 0.5°C in an incubator with 100 percent humidity.
- Count with the aid of a low-power microscope all dark colonies having a sheen or metallic-appearing surface luster, using for illumination a light source located directly above the filter.
- Calculation. The estimated coliform density of the sample is recorded as the number of coliform per 100 ml of sample.

$$\text{Coliform colonies}/100 \text{ ml} = \frac{\text{Coliform colonies counted} \times 100}{\text{ml sample filtered}}$$

The U.S. Public Health Service has established a standard of one coliform organism per 100 ml as the maximum allowable arithmetic mean density of all standard samples examined per month.

Standard Plate Count

This test provides an estimate of the total number of bacteria in a sample which will grow at 35°C in 24 hours under the conditions of food supply and moisture specified in the standard methods of analysis. Since many of the microorganisms which develop in this test are associated with animal life, the count can be an additional indication of pollution. The test is useful in judging the operating efficiency of various water treatment methods.

Use the same sample collected for the coliform test. The method of determination is as follows:

- Equipment
    - Glass petri dishes, 100-mm diameter, 15-mm depth
    - Porous covers for glass Petri dishes
    - Glass pipettes, straight-walled to deliver 1 ml
    - Erlenmeyer flasks, Pyrex, 125 ml

- Water bath, thermostatically set for 43-45°C
- Illuminated colony counter
- Reagents
  - Dehydrated Bacto tryptone glucose yeast agar
- Preliminary operations
  - Preparation of agar

    Weigh 12 gm of the dehydrated tryptone glucose yeast agar.

    Measure 500 ml of distilled water in a graduate.

    Pour 400 ml of the water into a beaker and heat to boiling.

    Suspend the 12 gm of agar in the remaining 100 ml of cold water.

    With constant stirring add the agar suspension to the 400 ml of boiling water. Continue stirring and boiling until the medium is completely dissolved.

    Pour equal amounts of the medium into each of ten 125-ml Erlenmeyer flasks.

    Close the Erlenmeyer flasks with cotton plugs.

    Sterilize in an autoclave for 15 minutes after the pressure has reached 15 lbs at a temperature of 121°C.

    Remove from the autoclave as soon as the pressure returns to zero pounds. The total time in the autoclave including heating, sterilizing, and cooling should not exceed 40 minutes.
  - Sterilization of glassware

    Wrap Petri dishes in groups of four in Kratt paper, or pack in metal cans.

    Wrap pipettes in Kraft paper or place in a metal pipette can.

    Sterilize in oven by heating at 170° C for 1 hour.
- Procedure
  - Melt the sterile agar by immersing the 125-ml Erlenmeyer flask in boiling water.
  - Cool the melted agar to 45°C and hold in a water bath at 43-45°C.
  - Shake sample violently in an up and down motion 25 times.
  - With a sterile pipette, transfer aseptically 1 ml of well shaken sample to a sterile Petri dish.
  - Add 10 ml of melted agar medium cooled to 43-45°C.
  - Uniformly mix and spread the medium and sample by rotating the plate on the surface of the work table.
  - Allow medium to harden and place Petri dish in incubator maintained at 35°C + 0.5°C.
  - After 24 hours, count the number of colonies on or in the agar medium using an illuminated colony counter.
  - If the colonies are too numerous to count, estimate the number by counting a fraction of the plate, such as one quarter or one tenth. Multiply this count by the appropriate factor.
- Calculation. Report results as "24-hour standard plate count per ml at 35°C."

116 Water Management and Supply

No definite standard can be set for the total count of untreated surface water. Untreated well waters should have a count of less than 500 per ml. Higher counts indicate possible contamination by surface drainage or by sewage. Chlorinated waters should have counts of less than 10 per ml. Higher counts indicate inadequate chlorination or recontamination after chlorination. In filtered water, high counts may indicate bacterial growths in filters, filtered water basins, or pipelines.

MICROSCOPIC EXAMINATION OF WATER

Taste and odor in water have many causes, one of the most important being the growth of aquatic microorganisms. The individual cells of these microorganisms are usually so small that they can be seen only with magnification. Storage reservoirs are particularly subject to such growths. Many of the microscopic plant forms belong to the algae group. Because algae contain chlorophyll and require sunlight for their growth and metabolism, most species are found only in open bodies of water, at or near the surface. Other species of microorganisms that do not contain chlorophyll grow and develop in the dark. They are found in covered reservoirs and at a depth in open reservoirs which sunlight will not penetrate.

Microorganisms Classifications

- Algae
  - Diatoms—unicellular plants with siliceous cell walls.
  - Cyanophyceae—true plants which contain chlorophyll and usually a pigment j sometimes called blue-green algae.
  - Chlrophyceae—chlorophyll-bearing plants with green pigment.
- Fungi
  Contain no chlorophyll, and therefore, do not require sunlight for growth. Bacteria belong to this class.
- Protozoa
  Single-cell animals requiring oxygen and organic food. Some are intermediate between the plant and animal kingdoms.
- Rotifera
  Minute animal forms.

All the many types of aquatic microorganisms secrete oils, which may be discharged during their life processes, and which are always set free after the death and disintegration of the cells. These oils are the cause of the tastes and odors imparted to water. Examples of tastes and odors caused by microorganisms are:

- Diatoms—aromatic and disagreeable.
- Cynophyceae and Chlorophyceae—vegetable, fishy, pigpen, disagreeable.

- Protozoa—disagreeable tastes, sometimes described as fishy, bitter, or cucumber.

In open bodies of water where conditions of light, food supply, and temperature are favorable, algae may grow sufficiently to produce a floating scum or "water bloom." Other microorganisms even grow under the ice during the winter. Growths of microorganisms, in addition to producing objectionable tastes and odors in water, may clog filters and reduce the length of filter runs. Fungi, as a rule, do not cause tastes but do produce slimes, generally in water mains. Many protozoa and rotifers are not taste producing when alive but are indicative of past pollution and a well-oxygenated water.

Clean glass or plastic bottles of at least 500-ml capacity should be used for the collection of water samples for microscopic examination. Samples should be collected at significant points representative of the entire body of water being examined and should correspond as nearly as possible to those points used in chemical and bacteriological studies in order to permit reasonable correlation.

In open reservoirs and other bodies of water, samples should be taken at a depth of 2 feet below the surface to detect chlorophyll-bearing species. Care should be taken to prevent contaminating the samples with floating scum growths. For the detection of nonchlorophyll-bearing species, samples should be collected at about a 20-foot depth.

For evaluation of the raw water received at the treatment plant, samples should be collected at the intake from the reservoir or other raw water source.

For the proper control of algae growths, enumeration of the specific species of microorganisms is essential. Identification of the predominant species will aid in determining the amounts of copper sulfate or other algicide needed to destroy the growths. Unless the growths in the sample to be examined are extremely heavy, concentration of the sample by filtration is required. This can be done with a Sedgewick-Rafter filter funnel using a small plug of standard graded sand as a filter. The microorganisms will be retained on the surface of the sand and may be washed out with a small quantity of water. In this way the microorganisms present in a large volume of sample can be concentrated to a final small volume.

- Equipment
  - Compound microscope providing magnification of 100 ×
  - Whipple ocular micrometer
  - Standard Sedgewick-Rafter counting cell with extra cover slips
  - Sedgewick-Rafter funnels
  - Wooden rack to hold Sedgewick-Rafter funnels
  - Silk bolting cloth discs, about 3/8 inch in diameter, having approximately 200 meshes per inch
  - Standard washed and graded sand for use in Sedgewick-Rafter method of filtration (this should be purchased ready for use)
  - Beakers, 50 ml

- Graduated cylinder, 10 ml
- Graduated cylinders, 500 ml
- Pipette, transfer 1 ml
- Rubber stoppers, one-hole #0
- Glass U-tubes to fit in rubber stoppers, with one arm about 1 inch longer than the other
- Wash bottle
- Preparation of filtering funnel
  - Fit a #0 one-hole rubber stopper with a small glass U-tube with the outer arm extending 1 inch above the small end of the stopper.
  - Place a silk bolting cloth disc moistened with distilled water over the opening in the small end of the rubber stopper.
  - Insert the rubber stopper with tube into the outlet end of the Sedgewick-Rafter funnel held vertically in a wooden rack.
  - Pour into the funnel sufficient standard sand to form a layer 1/2 inch deep on top of the disc.
  - Add 5-10 ml distilled water to wet the sand and remove entrapped air.
- Procedure
  - Measure 500 ml of well-mixed sample into a graduated cylinder.
  - Pour gently into Sedgewick-Rafter funnel, being careful not to disturb sand filter.
  - Allow sample to filter through sand, discarding effluent, until the water level is nearly down to the sand surface when filtration will automatically cease.
  - Remove stopper carefully, allowing sand and filter disc to fall directly into 50-ml beaker.
  - Using wash bottle, wash down the walls of the funnel with 4-5 ml of distilled water, allowing the wash water to run into the beaker.
  - Rotate beaker gently to permit water to wash, adhering microorganisms off the sand grains and filter disc.
  - Allow a moment for the course sand to settle, then decant the supernant promptly to a 10-ml graduate.
  - Repeat sand-washing process once or twice using small increments of distilled water, decanting each time into the 10-ml graduate.
  - Add distilled water to the graduate to build to volume and transfer to a clean 50-ml beaker.
  - Place a cover slip diagonally across the Sedgewick-Rafter counting cell.
  - Shake concentrated sample gently and withdraw 1 ml with pipette.
  - Introduce half at each open corner of the cell, and when this is carefully done, the cover slip will rotate automatically into position.
  - Place counting cell on stage of microscope and adjust tubes to give sharp focus, using a combination of lenses to provide magnification of 100X.
  - Using microscope, observe microorganisms. Compare size, shape, and characteristic markings of species observed in ten different fields with diagrams and photographs in standard reference books.

- Count any amorphous matter present in the fields examined.
- Calculation
  - Microorganisms are counted by surface area and not as individuals.
  - The standard unit for measuring surface area is a square, 20 microns on a side, or 400 square microns, known as the *areal standard unit*. This is the very smallest square marked off on the Whipple disc micrometer.
  - Using *standard units* square on an eyepiece micrometer as a guide, estimate and record the surface area covered by all individual microorganisms of each species noted in fields observed through a microscope.
  - Determine the factor needed to convert the values obtained in the concentrated sample to comparable values in the original sample.

$$\text{Factor} = \frac{\text{No. fields in 1 ml. counting cell}}{\text{No. fields counted}} \times \frac{\text{ml concentrate}}{\text{ml water in original sample}}$$

In the present case, where there are 1,000 fields in the counting cell, of which 10 are counted, and 500 ml of water is concentrated to 10 ml, the factor would be

$$\frac{1,000}{10} = X \frac{10}{500} = 2$$

- Determine the concentration of microorganisms in standard units per ml of original unconcentrated sample by multiplying the total areal units by the factor 2.

  Std units per ml = standards units in 10 fields × 2
- Interpretation of results include:

  - 100-300 areal standard units of microorganisms indicate active growth of algae.
  - 300-500 areal standard units may indicate possible taste and odor problems, or even shortened filter runs.
  - More than 500 areal standard units indicates a serious condition.
  - Less than 1,000 areal standard units of amorphous matter are not significant.
  - More than 1,000 areal standard units of amorphous matter indicate probable heavy growth of organisms that have died and disintegrated.
  - The presence of even one of certain microorganisms, such as Synura, is an indication of potential taste and odor conditions and the water should be promptly treated with an algicide such as copper sulfate.

5 WATER SUPPLY PURIFICATION

Processes through which the water quality is improved so that it is brought to conform or nearly conform to the quality standards (Table 5-1) is termed *water treatment*. Surface water contains impurities which are mostly colloidal in nature and, therefore, the conventional treatment of surface water may include:

- aeration
- sedimentation
- flocculation
- filtration
- disinfection

Groundwater, on the other hand, contains only dissolved impurities and, therefore, its treatment may be somewhat different. Natural waters contain many impurities, and modern, domestic, and industrial requirements are so rigorously established that it is necessary to purify natural waters. Systematic techniques have been developed for this purpose.

Volatile and oxidizable impurities are removed by aeration in fountain, spray, or cascade; suspended impurities, including much of the bacterial and microscopic contamination are removed by settling, coagulation, and filtration. Residual microorganisms are treated by chemical means or other forms of sterilization; soluble impurities (primarily of industrial importance) are removed by preheating, lime soda, or zeolite treatment.

Natural waters always contain dissolved gases such as oxygen, nitrogen, and carbon dioxide from the atmosphere. In addition, they frequently contain undesirable gases and odors which originated from decomposing organic matter in the water. Undesirable odors may usually be removed by purging the water with fresh air through the medium of natural stream flow or artificial aeration.

Table 5-1 Water Quality Standards

Substance or Characteristic	Maximum Acceptable Concentration	Maximum Allowable Concentration
Color on platinum cobalt scale	5	50
Taste	Unobjectionable	---
Odor	Unobjectionable	---
pH range	7.0 to 8.5	6.5 to 9.2
Total dissolved solids	500 mg/l	1500 mg/l
Turbidity in NTU (Nephalometer Turbidity Units)	5	25
Total hardness (as $CaCO_3$)	200 mg/l	600 mg/l
Calcium (as Ca)	75 mg/l	200 mg/l
Chloride (as Cl)	200 mg/l	600 mg/l
Copper (as Cu)	0.05 mg/l	1.5 mg/l
Fluorides (as F)	1.0 mg/l	1.5 mg/l
Iron (as Fe)	0.1 mg/l	150 mg/l
Magnesium (as Mg)	30 mg/l	150 mg/l
Manganese (as Mn)	0.005 mg/l	0.5 mg/l
Nitrate (as NO_3)	45 mg/l	45 mg/l
Sulphate (as SO_4)	200 mg/l	400 mg/l
Zinc (as Zn)	5.0 mg/l	15 mg/l
Anionic detergents (as MBAS)	0.2 mg/l	1.0 mg/l
Mineral oil	0.01 mg/l	0.30 mg/l
Phenolic compounds (as phenol)	0.001 mg/l	0.002 mg/l

Toxic substances

Arsenic (as As)	---	0.05 mg/l
Barium (as Ba)	---	1.0 mg/l
Cadmium (as Cd)	---	0.01 mg/l
Chromium (as hexavalent Cr)	---	0.05 mg/l
Cyanide (as CN)	---	0.05 mg/l
Lead (as Pb)	---	0.1 mg/l
Mercury (total as Hg)	---	0.001 mg/l
Selenium (as Se)	---	0.01 mg/l
Polynuclear aromatic hydrocarbons (PAH)	---	0.2 μg/l

Radioactivity levels

Gross alpha activity	---	3 pCi/l
Gross beta activity	---	30 pCi/l

Table 5-1 (continued)

Bacteriological levels

E Coli count in any 100-ml sample	---	0
Coliform organisms in 95% of 100-ml samples in any year	---	0
Coliform organisms in any 100-ml sample	---	10
Coliform organisms in second 100-ml sample, after it is found in first sample	---	0

[a]pCi = pico (10^{-12}) curie

AERATION

Aeration is a process of treatment whereby water is brought into intimate contact with air for:

- increasing the oxygen content
- reducing the carbon dioxide content
- removing hydrogen sulphide, methane, and various volatile organic compounds responsible for the taste and odors

Natural aeration takes place slowly in slow-flowing streams but rapid-flowing streams, especially with cascades or waterfalls, quickly absorb much oxygen from the air. This not only eliminates undesirable gases but at the same time oxidizes polluting substances such as ferrous iron, and others if they are present.

Artificial aeration is frequently used to supplement natural sources. This can be accomplished by causing the water to flow over cascades or baffles, thus spreading it in a thin layer, or by passing it through fountains, throwing it into a spray, or by running it through trickling fillers causing increased surface exposure. The last is also of special value for the oxidation of manganese and iron. Such methods frequently greatly improve the quality of natural water.

Where aeration is used to oxidize and precipitate iron, trickling can be used to good advantage. These are porous beds of broken stone, coarse sand, coke, shavings, or similar insoluble material having a large surface as compared to volume, through which the water slowly percolates. The air in such beds should be continuous or frequently changed or the process fails. Considerable attention has been paid to the design of spraying nozzles.

Aerators used for aeration purposes are normally classified as:

- waterfall aerators
- bubble aerators

Waterfall Aerators

Various types of waterfall aerators are:

- Multiple-tray aerator consists of four to eight trays, with perforated bottoms, at intervals of 300 to 500 mm. The rate of fall is 0.02 m^3/m^2 of tray surface. For a finer dispersion of water, the aerator trays can be filled with coarse gravel, about 100 mm deep.
- Cascade aerator consists of a flight of four to six steps each about 1 foot high with a capacity of about 3 fps per meter width. The cascade aerator is either circular or rectangular in area. A multiple-platform aerator uses the same principles. Sheets of falling water are formed for full exposure of the water to the air.
- Spray aerator consists of stationary nozzles connected to a distribution grid through which the water is sprayed into the surrounding air at a velocity of 15-20 fps.

Bubble Aerators

Bubble aerators include diffused plate and diffused tube aerators which are provided along one side of the length of a tank and suspended at mid-liquid depth. A venturi aerator, which is also a bubble aerator, requires a throat in the water pipe and a perforated air supply pipe. The amount of air required for bubble aeration is 3-5 ft^3 per ft^3 water.

The efficiency of an aerator is measured in terms of the percentage of CO_2 removed from water. The spray aerators are the most efficient.

SUSPENDED IMPURITIES

Suspended impurities, including turbidity from organic as well as inorganic matter, may be reduced by methods varying in efficiency, such as settling, coagulation, and filtration.

Settling

When water is impounded in reservoirs, suspended impurities gradually settle out, with the exception of living organisms which are floated by the gases they generate. When the reservoir can be constructed large enough and when microscopic growths do not develop, natural subsidence is frequently sufficient to produce clear water. The period necessary for accomplishing this result varies from a few hours to several days. To be efficient for such purposes, basins should be at least 8 feet deep to avoid scouring of the material from the bottom by wind-wave action.

Coagulation

When impounding alone may be insufficient or too slow, sedimentation may be hastened and improved by the addition of chemical agents. For this purpose coagulants are employed, such as salts of iron or aluminum, which will produce a coagulable precipitate of hydrate by reaction with the carbonate radical already present.

When such a precipitated hydroxide is formed in the water, it adsorbs and encloses suspended impurities including bacteria and causes these to settle much more readily. The reaction and coagulation, however, involve a time element and unless two hours or more are allowed for the precipitation, it may not be efficient or successful. Much longer retention in the coagulating basins is often found advantageous. Such coagulation basins are usually from 12 to 15 feet deep and the rate of flow through them is limited to 2.5 feet per minute. Baffle plates are often inserted to improve the mixing. The process usually removes 90 percent or more of the total turbidity and 75 percent of the bacteria.

Filtration

When settling alone is too slow or incomplete, filtration of the water is often resorted to. The usual filtering medium is sand and the method of application may be slow-sand, or rapid-sand, with or without pressure.

Sedimentation

Sedimentation is a process in which the suspended particles that are heavier than water are separated from water by gravitational settling. The terms *sedimentation* and *settling* are used interchangeably. A sedimentation basin may also be referred to as a sedimentation tank, settling basin, or a settling tank.

Depending on the concentration and the tendency of particles to interact, four types of settling can occur:

- Discrete particle settling
- Flocculant settling
- Hindered (also called zone) settling
- Compression settling

During a sedimentation operation it is common to have more than one type of settling occurring at the same time; and it is possible to have all four types occurring simultaneously.

Criteria for Sedimentation Tanks

The following criteria may be adopted for the design of sedimentation tanks:

- Depth of water in tank: 3-4.5 m (9-15 ft)
- Detention time: 1.5-4 hr (normally 3 hr)
- Number of units: 2 or more
- Velocity of flow: 0.3 m/min (1 ft/min)
- Surface loading: 1.5 m/h or 36 $m^3/m^2/d$ (average); maximum value 50 $m^3/m^2/d$
- Weir loading: 100-200 $m^3/m/d$
- Extra capacity for storage of sludge: 25 percent (manually cleaned)
- Floor slope: 1 in 12 or 8 percent, for mechanically scraped circular or square tank; 1 percent for mechanically scraped rectangular tanks (length/width > 2); 1.2:1-2:1 (vertical:horizontal) for hoppers without mechanical scraper
- Collecting weir should have: 90°, 50 mm deep V-notches at 150-300 mm apart; or circular orifices at 150-300 mm apart
- Velocity of water at outlet conduit: not more than 0.4 m/s
- Scraper velocity: One revolution in 45-80 minutes.

Sedimentation is gravitational separation of particulate matter from a suspension. It is the most widely used unit operation in water and wastewater treatment. The terms *sedimentation*, *settling*, and *clarification* are used interchangeably. Sedimentation is generally used in one or more steps of a wastewater treatment sequence, although in some cases it is the only treatment to which a wastewater is subjected. Based on the concentration and nature of the particles, the settling behavior of suspensions is defined by four types of settling:

- Type 1 or discrete particle settling
- Type 2 or flocculant settling
- Zone settling
- Compression settling

The four types of settling as encountered in wastewater treatment are presented in Table 5-2. A clear dividing line between different types of settling may not exist for a given application. During the process of sedimentation it is possible to have more than one type of settling occurring simultaneously.

Sedimentation tank design is based on the quantity of the water treated, the selected overflow rate, and the detention time. The overflow rate, also commonly referred to as the surface loading rate, is defined as the ratio of the flow rate to the surface area of the tank. For a given flow rate, the selected overflow rate establishes the surface area of the sedimentation tank. The tank depth is then calculated using the selected detention time.

The concentration of solids encountered in the physical chemical treatment of water and wastewater is generally not very high. Consequently, their settling behavior is represented by discrete particle settling. Theoretically, all particles having a settling velocity greater than the overflow rate will be completely

Table 5-2 Type of Settling Phenomenon Involved in Water and Wastewater Treatment

Type of Settling Phenomenon	Description	Application / Occurrence
Discrete particle (Type 1)	Refers to the sedimentation of particles in a suspension of low solids concentration. Particles settle as individual entities, and there is no significant interaction with neighboring particles.	Removes grit and sand.
Flocculant (Type 2)	Refers to a rather dilute suspension of particles that coalesce, or flocculate, during the sedimentation operation. By coalescing, the particles increase in mass and settle at a faster rate.	Removes a portion of the suspended solids in primary untreated primary wastewater in primary settling secondary settling facilities. Also removes chemical floc in settling tanks.
Hindered, also called zone (Type 3)	Refers to suspensions of intermediate concentration, in which interparticle forces are sufficient to hinder the settling of neighboring particles. The particles tend to remain in fixed positions with respect to each other, and the mass of particles settles as a unit. A solids-liquid interface develops at the top of the settling mass.	Occurs in secondary settling facilities used in conjunction with biological treatment facilities.
Compression (Type 4)	Refers to settling in which the particles are of such concentration that a structure is formed, and further settling can occur only by compression of the structure. Compression takes place from the weight of the particles, which are constantly being added to the structure by sedimentation from the supernatant liquid.	Usually occurs in the lower layers of a deep sludge mass, such as in the bottom of deep secondary settling facilities and in sludge-thickening facilities.

removed, regardless of the tank depth. Therefore, the settling velocity of the smallest particle whose complete removal is desired defines the overflow rate to be used in the design. In principle, tanks envisioned for the settlement of discrete particles should have large surface areas and shallow depths. The only reasons for using greater depths are:

- To provide space for sludge removal equipment
- To keep the horizontal flow-through velocity low enough so that particles which have settled will not be scoured or resuspended

In case of flocculating solids, the removal efficiency depends on depth. This is because increases in depth will increase the detention time, which in turn results in an increase in opportunities for particle-to-particle contact, leading to coalescence of such solids. This coalescence increases particle size and weight, helping the settling process. Additionally, unlike discrete particles, flocculating solids do not have a fixed density. Varying volumes of mother liquor (that is, water or wastewater) are incorporated into a settling floc as a result of collisions between particles. Flocs formed by coagulants can contain up to 99 percent water by volume and have extremely low densities.

Because of the complexity of the settlement of flocculating particles, the design overflow rate is generally established by laboratory batch settling tests. In these tests the suspension is placed in a column (typically 15 cm in diameter and 3 m in height with sampling ports located every 60 cm) and allowed to settle under quiescent conditions. The concentration of solids is determined on the samples withdrawn periodically from the sampling ports. The percentage removal of solids is calculated for each sample and recorded on a grid of depth and time of sampling. On this grid, lines of equal percentage removals, or isoconcentration lines are constructed. The resulting plot is used to determine the overflow rate and the detention time, by interpolation, for the desired percent removal of solids. Table 5-4 gives the recommended surface loading rates to be used in the design of sedimentation tanks for various suspensions. The detention times for such units range from 1.5 to 2.5 hours, with the weir loadings (that is, discharge per linear length of the effluent weirs) ranging from 125-500 m^3/m-day.

Settling tanks are classified according to the geometry of the horizontal cross-section. Sedimentation tanks in use today are either rectangular or circular in shape. A rectangular tank is shown in Figure 5-1. Settled solids in such tanks are led to a hopper normally located near the influent end by wooden flights mounted on endless chains. Circular tanks can be categorized into two general types: central feed and peripheral or rim feed (Figure 5-2). In the former, the wastewater is fed at the center where it enters a cylindrical well designed to distribute the flow equally in all directions. In the peripheral-feed type, a suspended circular baffle is located a short distance away from the tank wall so as to form an annular space into which the wastewater is discharged in a tangential direction. In both types of circular tanks the settled sludge is scraped into a hopper usually located

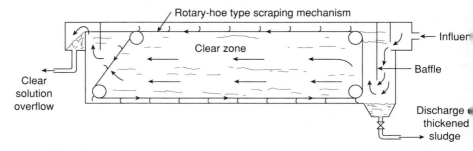

Figure 5-1 Rectangular clarifier

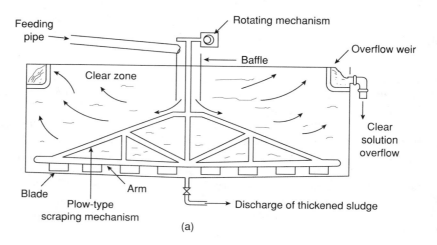

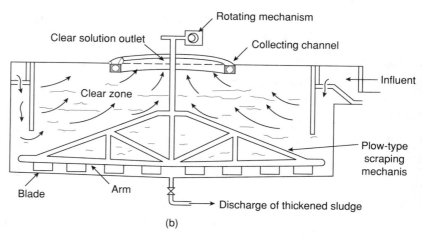

Figure 5-2 Circular clarifiers: (a) center feed; (b) peripheral feed

at the center. The sludge scraping mechanism is generally of the plow type, to overcome the inertia and adherence of sludge to the tank bottom. Typical dimensions used in both types of tanks are given in Table 5-3 for primary treatment of wastewater. Table 5-4 the shows recommended surface loading ratio for various chemical suspensions.

Table 5-3 Typical Design Information on Sedimentation Tanks Used for Primary Treatment of Wastewater

Item	Range	Value Typical	
Rectangular:			
Depth, m	3.0-5.0	3.6	
Length, m	15-90		25-40
Width, m	3-24		6-10
Flight travel speed, m/min	0.6-1.2	1.0	
Circular:			
Depth, m	3.0-5.0	4.5	
Diameter, m	3.6-60.0		12-45
Bottom slope, mm/m	60-160	80	
Flight travel speed, r/min	0.02-0.05		0.03

If widths of rectangular mechanically cleaned tanks are greater than 6 m (20 ft), multiple bays with individual cleaning equipment may be used, thus permitting tank widths up to 24 m (80 ft) or more.

Table 5-4 Recommended Surface-loading Rates for Various Chemical Suspensions

Suspension	Loading rate, $m^3/m^2 \cdot d$ Range	Peak Flow
Alum floc	25-50	50
Iron floc	25-50	50
Lime floc	30-60	60
Untreated wastewater	25-50	50

Mixed with settleable suspended solids in the untreated wastewater and colloidal or other suspended solids swept out by the floc.

Flotation

Solid particles may also be separated from liquids through flotation in which fine gas (usually air) bubbles are introduced in the liquids. These bubbles attach to the solid particles and will float.

Dissolved Air Flotation. In this system, air is dissolved in water under a pressure of several atmospheres, followed by the release of the pressure to the atmospheric pressure. In small systems, the entire flow can be pressurized by a pump with compressed air added at the pump suction. The entire flow is held in a retention tank under pressure for several minutes to allow time to dissolve the air. The solution is then admitted to a flotation tank through a pressure-reducing valve. The air comes out of the solution in the form of minute bubbles throughout the entire volume of the liquid. For large volumes, a portion of the water is recycled, pressurized, and partially saturated with air. The recycled flow is mixed with the unpressurized main stream prior to admission to the flotation tank.

Air Flotation. In this system, air bubbles are formed by introducing the gas phase into the liquid phase through a suitable device like a revolving impeller or through diffusers. Short-period aeration is not effective.

Vacuum Flotation. Initially water is saturated with air and later partial vacuum is formed. The dissolved air, in the form of minute bubbles along with attached solid particles, rises to the water surface.

Chemical Addition. Additives such as aluminum, ferric salts, and activated silica can be used to bind the particulate matter together which can easily entrap the air bubbles.

FLOCCULATION

Flocculation is a slow-mixing process in which destabilized colloidal particles are brought into intimate contact to promote agglomeration. The flocculation process depends upon physical factors such as time of flocculation, velocity gradient, concentration of particles per unit volume, and the size and nature of particles, and chemical factors such as alkalinity, pH, type of coagulant, and so on. Flocculation time is usually normally between 30 to 40 minutes, 30 minutes being the most common. For rapid mix, the time varies between 30 to 60 seconds, 60 seconds being the most common. The range of velocity gradient for rapid mix is 700-1,200 seconds. For flocculation which follows rapid mix, the range of velocity gradient is 10-80 seconds. The dimensionless product of time of flocculation and velocity gradient G should be in the range of 10^4 to 10^5. The concentration of the suspension is the turbidity present in water; the higher the turbidity, the more efficient is the process of flocculation.

Rapid-Mix Units

Rapid mixing is the first step in the water flocculation process and is carried out in rapid mix units, in which the destabilization of the colloidal particles is achieved. Contact time and velocity gradients are shown on Table 5-5.

Table 5-5 Contact Time and Velocity Gradient in Rapid Mixing

Contact time (seconds)	20	30	40	>40
Velocity gradient (per second)	1,000	900	790	700

The following types of rapid-mix units are used:

- Mechanical Mixer—This is the most commonly used type with a propeller impeller. The contact time in such a device reduces with an increase in the velocity gradient.
- Diffusers and Injection Devices—A rectangular grid of diffusers consisting of a series of tubes with orifices to develop the turbulence for mixing in downstream wakes is used. The coagulant is fed through the orifices.
- In-Line Blenders—This type of rapid mix can achieve instantaneous mixing; requires no head-loss computations and is low in cost since conventional rapid-mix facility can be omitted. It can generate velocity gradients in the range of 3,000-5,000 per second, and a residence time as small as 0.5 seconds is sufficient.
- Hydraulic Mixing—In this type of mixing, the flow is measured by a Parshall flume, followed by a hydraulic jump. Velocity gradient is about 800 per sec. The minimum residence time required is 2 seconds and requires no mechanical equipment.

The mixing can also be achieved through a recirculating pump. The delivery side has a nozzle operating under 24 ft of head.

Flocculators

A flocculator follows a rapid mix device. It helps in building up dense and large-size flocs, which in turn settle down quickly in the clarifier.

Several types of flocculators, depending upon their mixing devices, are used. Such flocculators and their mixing devices are described briefly. Several types of hydraulic flocculators are used. They are:

- Horizontal-Flow Baffled Flocculator—Figure 5-3 shows the plan of a typical horizontal-flow baffled flocculator. This flocculator consists of several around-the-end baffles with in-between spacing not less than 0.45 m to permit cleaning. Clear distance between the end of each baffle and the wall is about 1.5 times the distance between the baffles, but never less than 0.6 m. Water depth is not less than 1.0 m and the water velocity is in the range of 0.10-0.30 m/s. The detention time is between 15-20 minutes. This flocculator is suited for very small treatment plants being drained and cleaned. The head loss can be changed as required by altering the number of baffles.

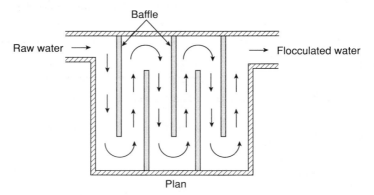

Figure 5-3 Horizontal-flow baffled flocculator in plan view

- Vertical-Flow Baffled Flocculator—Figure 5-4 shows the cross section of a typical vertical-flow baffled flocculator. The distance between the baffles is not less than 0.45 m. Clear space between the upper edge of the baffles and the water surface or the lower edge of the baffles and the basin bottom is about 1.5 times the distance between the baffles. Water depth varies between two to three times the distance between the baffles, and the water velocity is in the range of 0.1-0.2 m/s. The detention time is between 10-20 minutes. This flocculator is used for medium and large-size treatment plants.

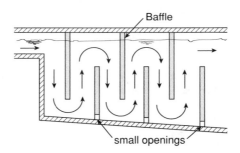

Figure 5-4 Vertical flow baffled flocculator (cross section)

- Hydraulic Jet Action Flocculator—This is a less-known type of hydraulic flocculator and is suitable for small treatment plants. In one type of jet flocculators, shown in Figure 5-5, the coagulant (alum) is injected in the raw water using a special orifice device at the inlet bottom of the tank. Water is then jetted into this hoppered tank. Helicoidal-flow (also called tangential-flow or spiral-flow) type flocculators as well as staircase-type flocculators can also be used.

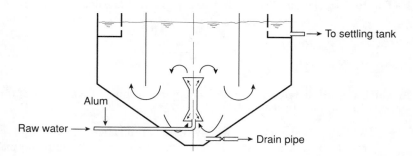

Figure 5-5 Jet flocculator

- Alabama-Type Flocculator—An Alabama-type flocculator, shown in Figure 5-6 is a hydraulic flocculator having separate chambers in series through which the water flows in two directions. Water flows from one chamber to another, entering each adjacent partition at the bottom end through outlets facing upwards. This flocculator was initially developed in Alabama and was later introduced elsewhere.

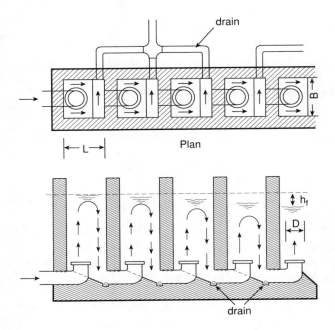

Figure 5-6 Alabama-type flocculator

The turbulence during the flow through a pipe can create velocity gradients leading to flocculation. The mean velocity gradient is calculated from

$$G = \left[\frac{\varrho g \, Q \, h_f}{\mu \, Vol} \right]^{1/2}$$

in which Q = flow rate (m³/S); Vol = volume of pipe of length L (m³); and h_f = head loss in pipe in length L (= $fL \, V^2/2gD$).

- Paddle flocculators are widely used in practice. The design criteria are—depth of tank: 3-4.5 m; detention time, t: 15-30 min, normally 30 min; velocity of flow: 0.2-0.8 m/s, normally 0.4 m/s; total area of paddles: 10-25 percent of the cross-sectional area of the tank; range of peripheral velocity of blades: 0.2-0.6 m/s, 0.3-0.4 m/s is recommended; range of velocity gradient, G: 10-75 s^{-1}; range of dimensionless factor Gt: 10^4-10^5; and power consumption: 0.06 to 0.08 kW/m³/min (0.09-0.12 kW/MLd).

For paddle flocculators, the velocity gradient is given by

$$G = \left[\frac{C_D \cdot A_p \varrho \, (V-v)^3}{2\mu \, Vol} \right]^{1/2}$$

in which C_D = coefficient of drag; A_p = area of paddle (m²); Vol = volume of water in the flocculator (m³); V = velocity of the tip of paddle (m/s); v = velocity of the water adjacent to the tip of paddle (m/s).

The value of C_D was assumed to be constant, but recent experiments have proved that it varies inversely with the value of G.

The optimum value of G can be calculated from

$$G_{opt}^{2.8} \, t \, C = 44 \times 10^5$$

in which G_{opt} = optimum velocity gradient s^{-1}; t = time of flocculation (min); and C = alum concentration (mg/l).

The shape of the container also affects the process of flocculation. For the same volume and height of water in the containers of several shapes such as circular, triangular, square, pentagonal, and hexagonal, it was observed that the pentagonal shape gave the performance of flocculation. (See Figure 5-7.)

- Pebble Bed Flocculator—The pebble bed flocculator contains pebbles ranging in size from 1 mm to 50 mm. The smaller the size of the pebbles, the better is the efficiency, but the faster is the build-up of the head loss; and vice versa. The depth of the flocculator is between 0.3 to 1.0 m. The velocity gradient is given by

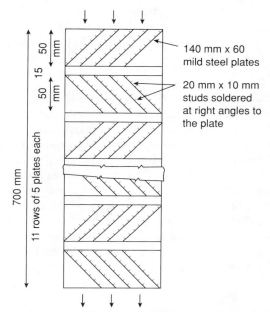

Figure 5-7 Surface contact flocculator

$$G = \left[\frac{\varrho\, g\, Q\, h_f}{\alpha\mu\, A\, L}\right]^{1/2}$$

in which h_f = head loss across the bed (m); α = porosity of bed; A = area of flocculator (m^2); and L = length of the bed (m). The main advantage of the pebble bed flocculator is that it requires no mechanical devices or electrical power. The operation and maintenance cost is also low and therefore it is useful for rural areas in developing countries. However, the drawback of this flocculator is that there is gradual build-up of the head loss across the pebble bed.

- Fluidized Bed Flocculator—In a fluidized bed flocculator, the sand bed is in the fluidized form. Even a 10 percent expansion of the sand bed is enough to create the required turbulence without choking the media. The sand size is between 0.2 to 0.6 mm. The flow of water is naturally upwards. This flocculator also does not require any mechanical equipment or electrical power. Further, there is no build-up of the head loss across the bed.
- Pneumatic Flocculator—In a pneumatic flocculator, air bubbles are allowed to rise through a suspension. This creates a velocity gradient useful for flocculation. The velocity gradient can be calculated from

$$G = 0.236 \frac{g\varrho D}{\mu} \left(\frac{Vol_A}{Vol}\right)^{1/2}$$

in which D = diameter of air bubbles and Vol_A/Vol = volume of air supplied per unit water volume. The flocculator needs compressed air and the problem of clogging of the diffuser is quite common. It is less efficient than the paddle flocculator and therefore less used in practice.

- In-line Flocculator—An in-line static flocculator or an in-line static mixer is a relatively recent device. It is housed in a gravity main and is static. The head loss in an in-line flocculator is also almost negligible. Only occasional flushing is necessary since deposition of some flocs takes place. The capitalized cost of a typical in-line flocculator is one third of the capitalized cost of conventional mixing impellers. Laboratory experiments show that twisted aluminum plates as static mixers in the pipeline gave better performance compared to the semicircular plates or the inclined-plane plates.

SLUDGE BLANKET CLARIFIER

A sludge blanket clarifier includes both flocculation and clarification as shown in Figure 5-7. Different problems involved in the conventional type clarifier are in connection with the doing and mixing, desludging, and the stability of the blanket. See Figures 5-8 and 5-9.

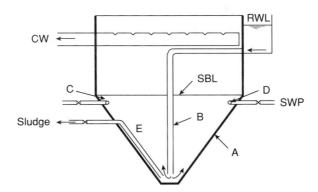

Figure 5-8 Candy-type upflow sludge blanket clarifier: RWL-raw water level; SBL-sludge blanket level; SWP-slurry withdrawal pipe CS-clarified water; A-upflow basin; B-raw water inlet pipe; C-slurry pocket; D-perforated slurry collection pipe; E-sediment withdrawal pipe

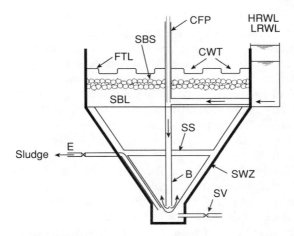

Figure 5-9 Sludge blanket clarifier; HRWL-high raw water level; LWRL-low raw water level; FTL-full tank level; SBL-sludge blanket level CFP-chemical feed pipe; CWT-clarified water troughs; SBS-sludge blanket stabilizer; SS-slurry slit; SWZ-sludge withdrawal zone; B-raw water; E-sludge withdrawal pipe; SV-scour valve

The velocity gradient of the sludge blanket can be calculated from

$$G = \left[\frac{g\varrho}{\mu}(S_S - 1)(1 - \alpha) \, h \left(\frac{C}{Q}\right)\right]^{1/2}$$

in which S_s = specific gravity of flocs; α = porosity of blanket, h = depth of blanket (m); C = capacity of clarifier (m^3); and Q = rate of flow (m^3/s).

- Tapered Velocity Gradient Flocculator

In a tapered velocity gradient flocculator, the water is initially subjected to a high-velocity gradient and finally to a low-velocity gradient, thus generating dense, large-size, and tough flocs which in turn settle more quickly. The efficiency of a tapered velocity gradient flocculator increases when there is an increase in the range of the velocity gradient, the mean value of G remaining the same. There is a gradual decrease in the velocity gradient and no sudden decrease along the direction of flow, and dual tapering in the velocity gradient as well as in the time of flocculation is achieved. That is, the highest velocity gradient for the shortest time is followed by a little lower value of G, the velocity gradient, for a little more time and so that in the end the value of the velocity gradient is the least with the maximum time of flocculation.

FILTRATION

Filtration is a physical process for the separation of nonsettling impurities and water or wastewater by passing through porous media. Filtration has been a principal unit operation in the treatment of potable water since the early nineteenth century. Interest in wastewater filtration has increased dramatically as a consequence of more stringent effluent standards. Wastewater filtration is considered a well-established process for the tertiary treatment of effluents from treatment plants. The purpose of tertiary wastewater filtration is to achieve supplemental removal of suspended solids, including BOD particulate from secondary effluents.

Filtration is carried out within a filter box consisting of a filtering medium and an underdrain collection system. Essentially, filters for treating potable water are similar to those employed for the treatment of wastewater. The processes have significant differences stemming mainly from influent characteristics. Solids present in secondary effluents, such as influent-to-wastewater filters, are larger, heavier, and have more variation in particle sizes. Filtration operations consist of two distinct phases: filtration and cleaning. As filtration progresses the solids removed within the filter medium accumulate, resulting in head loss build-up (increase in pressure drop) or loss of efficiency in removing suspended matter. The filtration is terminated when the limits for head loss or effluent suspended solids, usually referred to as *filter breakthrough*, are reached. Both of these events occur at the same time. Upon completion of the filtration phase, the filter bed requires cleaning before starting the next filtration cycle.

Although a wide variety of methods exists for filtration, including microstrainers and precoat filters, granular media filters are by far the most common method. Granular media filters can be classified as follows:

- The direction of flow: upflow or downflow filters
- The types of filter beds: single or multimedia filters
- The driving force: gravity or pressure filters
- The rate of filtration: constant or declining rate filters

Each of these filter types has its own merits in application. For example, multimedia filters were developed to achieve a more effective use of the filter bed depth. They allow the suspended solids to penetrate deeper into the bed and, therefore, use more of the solids-trapping capacity available within the filter. In single medium filters most of the solids removal is realized in the surface few centimeters of the bed. Deeper penetration of solids results in a decrease in the rate of head loss accumulation and leads to longer filter runs. Granular media filters are further classified with respect to the magnitude of the rate of filtration as rapid or slow. (See Figure 5-10.) This seemingly small difference results in a different filtration behavior and methods of cleaning.

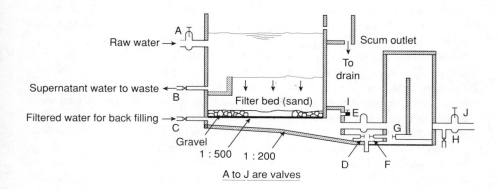

Figure 5-10 Schematic section of a slow-sand filter

Rapid Filters

Most rapid filters operate in the downflow mode and are gravity systems. Gravity filters are usually built with open tops and operate with an available head of 2 to 4 m (6.5-12 ft). Pressure filters operate in the same manner as gravity filters and are generally suited for smaller plants. The only difference between them is that the filtration operation in pressure filters is carried out in a closed vessel, usually in the form of a cylindrical tank, under pressurized conditions achieved by influent pumping. The operational head for pressure filters may be 35 to 100 m. Upon reaching the maximum allowable head loss or filter breakthrough, rapid-sand filters are cleaned by backwashing. Backwashing is the operation of passing water in the upward direction at a high enough rate to fluidize the bed so as to flush out the entrapped solids.

Granular media commonly used in wastewater filters are:

- Silica sand, specific gravity of 2.65
- Anthracite coal, specific gravity of 1.35-1.75
- Garnet, specific gravity of 4.1-4.6

Proper design of a filtration unit should cover details regarding size, uniformity, and depth specifications for the media. Most commonly, media specifications are reported in terms of the effective size and the uniformity coefficient, which are determined by standard sieve analyses. The effective size is defined as the diameter for which 10 percent of the sand is finer by weight. On the other hand, the uniformity coefficient is defined as the ratio of diameter (for which 60 percent of the sand is finer by weight) to effective size. Generally, low uniformity coefficients, for example, less than 2.0, are desirable, or higher backwash rates would be required to fluidize coarser grains. Another design parameter is the rate of filtration. Normal filtration ranges are 80-200 $1/m^2$-minute at average flow rates, which may be allowed to increase up to 400 $1/m^2$-minute for peak flows. Higher filtration rates require less filter area, for a given rate of flow, they have shorter filter runs due to the increase in solids loading and require more frequent backwashing. It is recommended that the length of filter runs should be 15 to 30 hours at a filtration rate of 200 $1/m^2$-minute. This corresponds to filtering about 200 to 400 m^3 of wastewater per m^2 of filter area per filtration cycle. Typical operating conditions for filtration of wastewaters are summarized in Table 5-6.

Completion of the filtration phase is followed by the cleaning of the filter bed by backwashing. The flow rates that bring about fluidization in a multimedia bed are not the same for each layer; for example, a backwash rate of 600 $1/m^2$-minute would be satisfactory for filter beds of 0.4-0.6 mm sand. Such a rate is expected to lead to a 40-50 percent bed expansion which would lead to a high-hydraulic shearing force on the sand grains. Larger and/or heavier media may necessitate backwash rates as high as 2,000 $1/m^2$-minute. When comparing the rapid filtration of potable water with wastewater, two points are worth mentioning regarding the backwashing operation. Backwashing alone may not result in satisfactory cleaning of wastewater filters, because of the biological solids in some effluents which can strongly attach to the medium due to its sticky nature. Auxiliary operations such as surface washing or air scouring may be needed to enhance the scouring action due to normal backwashing. Quantities of backwash water produced in wastewater filters are much higher in wastewaters than in potable water filters, for example, 5 percent to 10 percent compared to 2 percent to 5 percent of the volume filtered. This is due to the shorter filter runs and more frequent backwashing because of the higher solids loading involved. Since the backwash water is retreated in the plant, it produces an added burden on plant operations. Figure 5-11 shows a schematic of a rapid-sand filter.

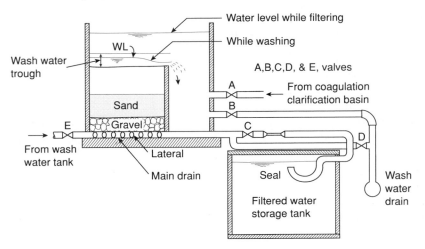

Figure 5-11 Rapid-sand filter

Table 5-6 Typical Operating Conditions for Filters

Characteristic	Value Range	Typical
Single medium (shallow bed):		
Sand:		
Depth, mm	25-40	35
Effective size, mm	0.5-1.0	0.8
Uniformity coefficient	<2	1.6
Filtration rate, $1/m^2$-minute	<200	120
Single medium (deep bed):		
Sand or anthracite:		
Depth, mm	1,000-2,000	1,200
Effective size, mm	1-3	1.5
Uniformity coefficient	<3	2.0
Filtration rate, $1/m^2$-minute	>200	400
Dual medium:		
Anthracite:		
Depth, mm	300-600	450
Effective size, mm	0.8-2.0	1.2
Uniformity coefficient	1.3-1.8	1.6
Sand:		
Depth, mm	150-300	300
Effective size, mm	0.4-0.8	0.55
Uniformity coefficient	1.2-1.6	1.5
Filtration rate, $1/m^2$-minute	80-400	200
Multimedium:		
Anthracite (top layer of quadmedium filter):		
Depth, mm	200-400	200
Effective size, mm	1.3-2.0	1.6
Uniformity coefficient	1.5-1.8	1.6
Anthracite (second layer of quadmedium filter):		
Depth, mm	100-400	200
Effective size, mm	1.0-1.6	1.2
Uniformity coefficient	1.5-1.8	1.6
Anthracite (top layer of trimedium filter):		
Depth, mm	200-500	400
Effective size, mm	1.0-2.0	1.4
Uniformity coefficient	1.4-1.8	1.6
Sand:		
Depth, mm	200-400	250
Effective size, mm	0.4-0.8	0.5
Uniformity coefficient	1.3-1.8	1.6

Table 5-6 (continued)

Characteristic	Value Range	Value Typical
Garnet or ilmenite:		
Depth, mm	50-150	100
Effective size, mm	0.2-0.6	0.3
Uniformity coefficient	1.5-1.8	1.6
Filtration rate, $1/m^2$-minute	80-400	200

Slow-Sand Filters

The slow filters are thus called because the rates of filtration employed are 15 to 150 times less than those of rapid filters and, therefore, their area requirements are correspondingly larger. The granular medium used in slow filters is almost always sand. As in the case of rapid filters, slow-sand filters are traditionally used for potable water treatment. In spite of the inherent disadvantage of large area requirement, slow-sand filters have been and are still being used extensively. They offer the following advantages over the rapids and filters:

- Stable and effective suspended solids and bacterial removal.
- Low cost of construction.
- Simplicity in design, that is, little pipework and instrumentation, leading to minimal operational problems and skilled labor requirements.
- In particular, they are noted for their effectiveness in bacterial and viral removal with reductions over 99.9 percent.

During the operation of slow-sand filters, a layer of inert deposits and biological material forms on the surface of the sand bed. This layer is referred to as the *schmutzdecke*, a German word meaning dirt layer. Biological growth also occurs within the sand bed; both schmutzdecke and biological growth within the filter play important roles in the effectiveness of slow-sand filters and may take weeks or months to develop. The cleaning process of slow-sand filters is different than that of rapid filters. It covers the removal of schmutzdecke by scraping 2-4 cm from the surface on completion of the filtration phase. Removed sand can be washed and either replaced or stored for future use. In the latter case, a gradual decrease in the sand bed depth takes place, and the process is continued until the minimum sand bed depth (usually 50 to 60 cm) is reached. The frequency of cleaning depends on the influent characteristics and the rate of filtration and can be on the order of several weeks to months.

Rapid-sand or slow-sand filtration as applied to wastewater effluents has no generalized approach for the design of full-scale units. This is due to the inherent variability in the characteristics of effluents. Consequently, for a given application, pilot plant studies are required to ensure that the selected filter configuration will function as desired. Experience has indicated that the performance of a pilot filter column with a diameter of at least 15 cm can simulate operating results from a full-scale filter. For the case of rapid filters a minimum cross-sectional area of 0.1 m^2 is suggested to accurately determine backwashing characteristics. The influent water system should be able to provide the filtration rates typically employed under field conditions. Pilot units should also allow observation of the head loss development up to the maximum level expected for the termination of the filtration phase. Filter columns should be connected in parallel to make sure that they receive the same wastewater as their influent. It is recommended that there should be as many pilot filter columns as there are variables to be considered. Testing should be carried out for a full year to take seasonal variations in the influent characteristics into account.

FILTER MEDIA

Sand has been used traditionally as a filter medium because it is easily available, inert, hard, durable, almost spheroid in shape, and heavier (sp. gr. 2.65) than water. The effective size (the size corresponding to 10 percent passing through) of sand is in the range of 0.45-0.55 mm, 0.45 mm being the most common. The depth of sand should not be less than 0.6 m and is generally not more than 0.9 m. Average depth of sand is 0.75 m.

The gravel bed surrounds the underdrain system, supports the sand bed above it, and also contributes to the uniform distribution of the backwash water. Gravel should preferably be rounded pebbles and not crushed stone. A gravel bed is not needed when the filter is equipped with porous filter plates which directly support the sand bed as indicated in Table 5-7.

Table 5-7 Sizes and Depths of Gravel Layers for Rapid Sand Filters

Range in Size mm	Range in Depth mm
65-38	130-200
38-20	80-130
20-12	80-130
12-5	50-80
5-2	50-80

The underdrain system collects the filtrate and carries it to the clear water storage and also distributes the wash water uniformly to the filter bed during the backwash operations. It usually consists of perforated pipe laterals and a centrally located manifold. It is therefore known as *lateral and manifold* system.

For uniform backwashing of the filter, the discharge through the orifices should almost be the same, whatever their distance from the manifold. For this purpose, the allowable friction loss (h), in a lateral with (n) orifices must equal ($1-m^2$) times the driving head (h_d) on the first orifice reached by the wash water if the head on the nth orifice is not to exceed m^2 times the driving head, where m is almost equal to unity, or 0.9.

With the lost head and neglecting the side effects, the friction in the perforated lateral of unvarying size must be numerically equal to the friction exerted by the full incoming flow in passing through one third the length of the lateral. For an optimal hydraulic system, 25 percent of the overall head loss should occur in the delivery of the required flows to the orifices, and the remaining 75 percent should be for driving the wash water, in succession, through the orifices, gravel, and fluidized bed.

The rule of thumb for the design of the underdrain system for filters washed at the rate of 0.15-0.90 m/min is:

- Ratio of the area of orifice to area of bed served—(1.5-5.0) x 10^{-3}.1
- Ratio of the area of laterals to area of orifices served—(2-4):1
- Ratio of area of manifold to area of laterals served—(1.5-3):1
- Diameter of orifices: 6-18 mm
- Spacing of orifices: 75 mm for 6-mm diameter, 300 mm for 18-mm diameter
- Spacing of laterals: closely approximating spacing of orifices
- Length of laterals on each side of manifold—not more than 60 times their diameter
- The orifices are oriented so that wash water is directed downward at an angle of 30°-60° with vertical corrosion or corrosion of the metal around these holes may be minimized by lining the holes with a brass or bronze bushing.

Filter floors, also known as false bottoms or false floors, may be used to replace manifold and lateral systems. They support the filter sand bed possibly without stone and gravel in transitional layers below the filter bed, creating a single box-like waterway beneath the filter to collect the filtered water or to provide the wash water. The floor is perforated by short tubes or orifices for the even distribution of the wash water.

There are two methods of filter backwash: backwash with water alone and hard wash. An air-water wash system in which air agitation followed by water wash is adopted.

In hard wash, the rate of backwash for 50 percent expansion is about 72 m/h. A second system, free air at the rate of 0.9-1.5 m^3/m^2 of filter area/minute, is forced through the underdrains for a period of 5 minutes for agitation of the sand. Wash water is then introduced through the same underdrains at the rate of 19-29 m^3/m^2 of filter area/hour.

Wastewaters

An auxiliary scour is available using a surface wash system used in addition to the conventional system of backwash. The primary pipe grid system consists of horizontal pipes suspended on wash water troughs. The secondary vertical pipes are connected to horizontal headers. The vertical pipes are extended to within about 100 mm of the surface of the unfluidized bed. The bottom end of the vertical pipes are 0.6-0.75 m c/c and work under a head of 7-20 m of water. A filter bed agitator which consists of a horizontal revolving pipe supported at its center is used. Several nozzles are attached to this pipe. Such units, either one or more, are located at a height of 25 mm above the unexpanded sand surface. The nozzles face downward at 30° to the horizontal. The resulting jet action causes the arm to revolve up to 20 rpm depending upon the length of the pipe. Mechanical revolving rakes are forced through the sand bed. The normal speed of the rake is 10-12 rpm.

Wash water troughs are located above the 50 percent expanded sand level to collect the wash water and discharge it into a drain pipe. The maximum horizontal travel of wash water should not exceed 0.9 m and therefore the spacing between the troughs should not exceed 1.8 m. The free tall discharge through the gutter is given by $Q = 1.375\ b\ h_o^{3/2}$ in which Q = rate of discharge (m³/s); b = width of the gutter (m); and h_o = initial depth of water (m).

The capacity of a wash water tank should be 1 percent to 6 percent of the filtered water and should be sufficient for at least 10 minutes wash of one filter or 5-6 minutes wash of two filters without refilling. The bottom of the tank is normally located about 12 m above the underdrain system. The wash water head should be around 4.5 m just outside the underdrain system.

Filter accessories include manually, hydraulically, pneumatically, or electrically operated gates on the influent, effluent, drain, and wash water lines; measuring and rate-control devices; head-loss gauges; and wash water controllers and indicators. Among such equipment is an automatic effluent-flow regulator. The valve mechanism automatically opens or closes to keep the discharge rate constant. Pressure differentials between the venturi throat and outlet are translated into valve movements by a balancing diaphragm or a piston. Problems associated with rapid sand filters are:

- air binding
- mud accumulation
- sand incrustation
- negative head

Table 5-8 is a comparison of conventional slow-sand and rapid-sand filters.

Table 5-8 Comparsion of Conventional Slow-Sand and Rapid-Sand Filters

Characteristic	Slow-Sand Filters	Rapid-Sand Filters
Rate of filtration	0.1-0.4 m/h (0.2 m/h)[a]	4-20 m/h (5 m/h)
Size of bed	Large, 2,000 m^2	Small 40-400 m^2
Depth of bed	0.3 m of gravel, 0.9-1.1 m sand, usually reduced to not less than 0.5-0.8m by scraping	0.3-0.45 m of gravel, 0.6-0.9 m of sand, not reduced by washing
Size of sand	Effective size: 0.20-0.30 mm Uniformity coefficient: 2-3.5 (2.5)	Effective size: ≥ 0.45 mm Uniformity coefficient: ≤ 1.5, depending upon underdrainage system
Grain size distribution of sand in filter	Unstratified	Stratified with smallest or lightest grains at top and coarsest or heaviest at bottom
Underdrainage system	1. Split tile laterals laid in coarse stone and discharging into tile or concrete main drains 2. Perforated pipe laterals discharging into pipe mains	1. Perforated pipe laterals discharging into pipe mains 2. False floor type, with orifices 3. Many others, generally proprietary
Loss of head	60 mm initial to 1.2 m final	0.3 mm initial to 2.4-2.75 m final
Length of run between cleanings	20-60d (30 d)	12-72 h (24 h)
Penetration of suspended matter	Superficial	Deep, particularly with dual or mixed media
Method of cleaning	Scraping of surface layer of sand and washing and storing cleaned sand for periodic resanding of bed	Dislodging and removing suspended matter by upward flow or backwashing which fluidizes the bed Possible use of auxiliary scour system

[a] Common value shown within parentheses

Table 5-8 Comparsion of Conventional Slow-Sand and Rapid-Sand Filters *(Continued)*

Characteristic	Slow-Sand Filters	Rapid-Sand Filters
Amount of water used in cleaning sand	0.2-0.6% of water filtered	1-6% (4%) of water filtered
Preparatory treatment of water	Generally none when raw water turbidity < 50 NTU	Coagulation, flocculation and sedimentation
Supplementary treatment of water	Chlorination	Chlorination
Cost of construction	Relatively high	Relatively low
Cost of operation	Relatively low where sand is cleaned in place	Relatively high
Depreciation cost	Relatively low	Relatively high

CLASSIFICATIONS

Rapid-sand filters can be classified in several ways. They may be classified according to

- type of filter media employed
- type of filter rate-control system employed
- direction of flow through the bed
- whether they operate under gravity (free surface) or pressure, that is whether they operate under atmospheric or more than atmospheric pressure

In the conventional rapid-sand filters, the pore size of the filter bed is the smallest in the topmost layers. Therefore, the floc particles removed during the filtration process concentrate in these layers and most of the lower layers remain unused. This has led to the introduction of dual and multimedia filters, in which lighter media of larger size occupy the upper layers. Such flocs can penetrate deeper and lower layers of the filter become effective.

A dual-media filter (Figure 5-12) is generally composed of a coarse coal upper layer (sp.gr.: 1.45-1.55; effective size: 1.0-1.6 mm; uniformity coefficient: 1.3-1.7; and depth: 0.40-0.75 m) on top of a lower sand layer (sp.gr.: 2.65; effective size: 0.45-0.80 mm; uniformity coefficient: 1.3-1.7; and depth: 0.15-0.30 m). As the specific gravities of the two materials are different, the two layers retain their respective positions after backwashing, although some intermixing of the layers occur at the interface.

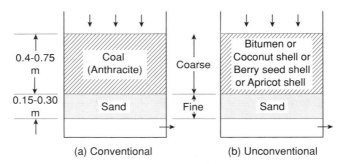

Figure 5-12 Dual-media filters

The advantages of the dual-media filters over the conventional filters are:

- Higher filtration rates (10-15 m/h) and therefore the reduction in the total filter area and cost for a given design capacity.
- Longer filter run at any given loading.
- Feasibility of increasing, at low cost, the capacity of the existing filters by their conversion to dual-media filters. Anthracite is the most accepted coarse medium.

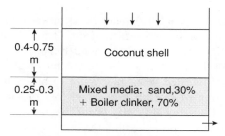

Figure 5-13 Dual-media

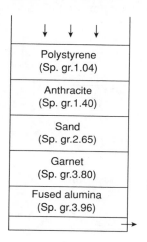

Figure 5-14 Multimedia filter

In multimedia filters (Figures 5-13 and 5-14), five media have been used: polystyrene (sp.gr.: 1.04), anthracite (sp.gr.: 1.4), crushed flint sand (sp.gr.: 2.65), garnet (sp.gr.: 3.8), and fused alumina (sp.gr.: 3.96). A constant-rate filter is used. Throttling valves are controls used to measure flow and differential pressure, or water-level devices are used to control the rate at the inlet or the outlet. The simplest method of flow control is to distribute the raw water by splitting the flow equally over the filter units. The water level rises during filtration to compensate for the head loss build-up in the filter bed. When the water level reaches the maximum permissible level above the filter bed, the filter unit is taken out for backwashing. When it is desirable to maintain a constant water surface in the filters, a rate controller must be provided at the outlet. A simple control device consists of a float connected by a small cable running over sheaves to a butterfly valve on the filter outlet pipe. When no rate controllers are used and the filter inlet is placed below the minimum water level, during filtration as the head loss increases the filtration rate decreases. However, the breakthrough of particles is less likely at the end of the run compared to the former type. The conventional rapid sand filter is an example of a downflow filter.

In an upflow filter (Figure 5-15) water flows upward, under pressure, through coarse-to-fine media. As the entire filter depth (1.8-3.0 m) is utilized, the head loss is low and the filter runs are typically longer. The rate controller and negative head are absent. As the filter acts as a flocculator at the inlet zone, a separate flocculator can be eliminated when raw water turbidity is less than 200 NTU. Because of fluidization of the top fine layer, loss of sand and the consequent deterioration of the filtrate quality may occur. To prevent this, a grid consisting of parallel vertical plates is placed within the bed and slightly below the top of the media. The grid spacing is 100-150 times the size of the fine sand. Formation of the inverted arches of sand minimizes the fluidization of sand. The rate of filtration can be 15-30 $m^3/m^2/h$.

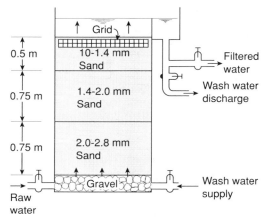

Figure 5-15 Single-medium upflow filter

The problem of the fluidization of the upper layer in an upflow filter is solved in a biflow filter, as shown in Figure 5-16. Here the effluent-collecting pipe is placed in the upper layers of the sand bed and the filtration take place simultaneously—from above, through the top layer of the sand; and from below, through the bulk of the media. The effluent pipe is located at a depth of about 20 percent of the total depth of the filter. The depth of sand is 1.50-1.80 m. The total filtration rate can be as high as 18 $m^3/m^2/h$, the downward rate being one half to two thirds the upward rate. The backwash rate is 54 $m^3/m^2/h$ for a period of 6 minutes.

In a radial flow filter, or continuous sand filter, water passes radially through a gradually descending column of sand (Figure 5-17). The impurities are removed continuously and there are no stoppages for backwashing. Chemically treated water enters into a central hollow column, permeates radially through sand, reaches the peripheral ducts, and flows out of the shell. The dirty sand is continuously drawn from the bottom, washed and lifted with air, and placed again at the top. The sludge separated from the sand is withdrawn at the top.

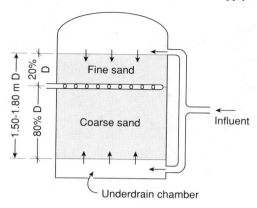

Figure 5-16 Biflow filter

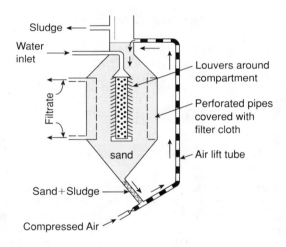

Figure 5-17 Radial flow filter

The decline velocity filter (Figure 5-18) is placed in a trapezoidal container and the direction of flow is from the bottom to the top. As the water travels upward, the velocity of water decreases giving a better opportunity for the suspended particles to be removed in the filter bed. Thus, this filter combines the features of upflow and radial flow filters.

Several inlets and outlets are provided in a multi-inlet, multi-outlet filter (Figure 5-19) to give parallel filter units, one over the other. Uniform sand with uniformity coefficient 1.2 and 0.45-0.60 mm size is used. The effective depth of each filter layer is about 0.2-0.3 m. The location of the outlet pipe corresponding

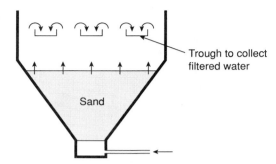

Figure 5-18 Decline velocity filter

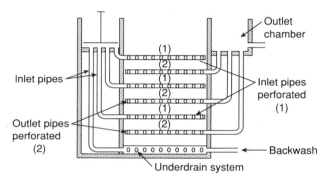

Figure 5-19 Multi inlet multi outlet filter

to the inlet pipe is above the inlet pipe and therefore negative head is not developed and mud balls are also absent. The effective filtration rate is only 2-3 $m^3/m^2/h$ but the yield is higher per unit occupied area.

Pressure filters work on the same principle as gravity type rapid-sand filters, but the water is passed through the filter under pressure. The pressure filter (Figure 5-20) is housed in a cylindrical tank usually made of steel or cast iron. The tank axis may be either vertical or horizontal. Pressure filters are compact and can be prefabricated and moved on site already assembled. Coagulation and flocculation are achieved in the top portion of the filter. Economy is further possible by avoiding double pumping. These filters are most suitable for small applications but the effectiveness of the backwash cannot be directly observed; and it is also difficult to inspect, clean, and replace the sand, gravel, and underdrains.

Direct filters work on the principle of direct filtration, which involves unit processes incorporating coagulation, flocculation, and filtration without the use of sedimentation. Direct filters can be used in a variety of applications. Direct filtration reduces alum usage and sludge production. The process of direct filtration can consistently produce high-quality filtered water with significantly low annual operating cost. At a filtration rate of 1.4-4 mm/s, direct filtration can reduce turbidity of 1-100 NTU to 0.1-0.3 NTU, comparable to that produced by

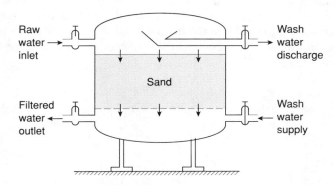

Figure 5-20 Pressure filter (horizontal type)

conventional treatment. Process control is also somewhat easier than with conventional treatment.

A diatomaceous filter (not strictly a rapid-sand filter but comparable in performance) has a medium consisting of diatomaceous earths which are skeletons of diatoms mined from such deposits. The filtering medium is a layer of diatomaceous earth built up on a porous septum by recirculating a slurry of diatomaceous earth until a firm layer is formed on the septum. The precoat thus formed is used for straining the turbidity in water. For this, diatomaceous earth is applied at 0.5 2.5 kg/m^2 of the septum. Sometimes, when the turbidity is very high, diatomaceous earth is added to the incoming water as body feed. Diatomaceous filters are of two types pressure type; and vacuum type (Figure 5-21).

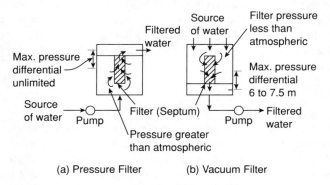

Figure 5-21 Diatomaceous filters

Valveless, self-washing filters with automatic siphon (Figure 5-22) work independently and automatically, without the use of electricity or auxiliary fluid systems, during both the filtration and washing cycles. The water in the buffer tank, after filtration, rises from the bottom of the filter to the filtrate storage tank at the top. When the storage tank is full, the filtrate is supplied through the outlet. When the filter bed is clogged, the water level rises in the buffer tank and also in the upstream arm of the siphon. As the maximum design head loss is reached, the siphon is primed and the water in the filtered water storage tank flows back, thus backwashing the filter. The capacity of the storage tank is designed to ensure adequate washing of the filter.

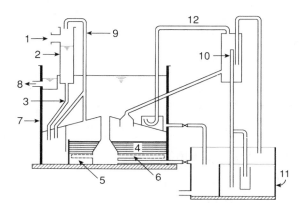

Figure 5-22 Valveless self-washing filter: (1) raw water inlet; (2) header tank; (3) raw water inlet pipe; (4) filter layer; (5) supporting layer; (6) filtrate pipe; (7) filtrate storage tank; (8) filtrate outlet; (9) vent pipe; (10) wash water siphon; (11) wash water return; (12) unpriming pipe

Backwashing takes place automatically at a predetermined fixed head loss and therefore the filter bed is never excessively clogged. Furthermore, compressed air and electrical power are not required; however, the filtrate supply is not available during the period in which the storage tank is being refilled. It is suitable for waters with low or medium suspended-solids content and the maximum filtration rate is about 10 $m^3/m^2/hr$.

Slow-Sand Filters

The main features and some design criteria of a slow-sand filter are already given in Table 5-8. The maximum turbidity that a slow sand filter can handle is 100-200 NTU for a few days, whereas 50 NTU is the maximum that should be for longer periods. Surface waters can be treated by slow-sand filters when turbidity is brought down to this level through flocculation and sedimentation or, in addition to this, through roughing filters. Turbidity can also be lowered by a presedimentation tank.

Roughing filters are used for pretreatment rather than for the treatment itself. In the case of high turbidity, roughing filters should precede slow sand filters in general, or occasionally rapid-sand filters. In an upflow filter (Figure 5-15), three layers having grain sizes from the bottom upward with a simple underdrain system would serve as roughing filter. Such a filter would have large pores and hence gradual clogging. The filtration rate may be up to 20 m/h. However, the time for backwash is relatively longer, for example, about 20-30 minutes.

A roughing filter can also be horizontal (Figure 5-23), having a depth of 1-2 m. The horizontal water flow rate, computed over the full depth, would be 0.5-1.0 m/h, giving a surface loading of only 0.03-0.1 m/h. The clogging is very gradual.

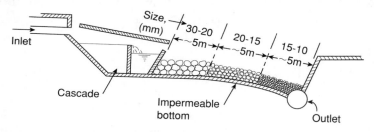

Figure 5-23 Horizontal roughing filter

DISINFECTION

Processes like flocculation, sedimentation, and filtration render the water chemically and physically or aesthetically acceptable with a slight reduction in the bacterial content. However, they do not provide completely safe water and therefore it is necessary to disinfect the water to destroy, or inactivate, disease-producing organisms. According to Chick's law, the number of organisms destroyed in unit time is proportional to the number of organisms remaining. Thus,

$$\ln \frac{N_t}{N_o} = -k^{+1/2} t^m$$

in which N_o = initial number of organisms; N_t = number of organisms remaining after time t; k = rate constant; t = time; and m = constant (> 1, < 1, $= 1$ when rate of kill rises in time, falls in time, remains constant, respectively).

The concentration of disinfectant is given by

$$C^n t_p = \text{constant}$$

in which C = concentration of disinfectant: t_p = time required to effect a constant percentage, usually 99 percent destruction of organisms; and n = a coefficient of dilution or a measure of order of reaction = 0.86 for most of the organisms).

The concentration of organisms can be expressed as

$$C^q N_p = \text{constant}$$

in which C = concentration of disinfectant; N_p = concentration of organisms reduced by a given percentage usually 99 percent in a given time; and q = coefficient of disinfectant strength.

The effect of temperature is given by

$$\log \frac{t_1}{t_2} = \frac{E(T_2 - T_1)}{4.56 \, T_1 T_2}$$

in which t_1, t_2 = times required for equal percentage of kill at fixed concentrations of disinfectant; t_1, t_2 absolute temperatures for which the rates are to be compared; and E = activation energy (calories).

Types of Disinfection

Physical disinfection can be obtained by the following methods.

- Heat—Boiling is a safe and time-honored practice that destroys pathogenic micro-organisms. It is effective as a household treatment but impractical for community water supplies.
- Light—Sunlight is a natural disinfectant, principally as a desiccant. Irradiation by ultraviolet light (UV) intensifies disinfection. It involves the exposition of a film of water, about 120-mm thick, to one or several quartz mercury vapor lamps emitting invisible light. In this method, the exposure is for short periods, no foreign matter is introduced, no taste and odor are produced; and overexposure does not result in any harmful effects. Effectiveness is significantly reduced when the water is turbid or contains nitrates, sulphates, and ferrous ions. No residual that would protect the water against new contamination is produced. Ultrasonic waves at 20-400 kHz frequency provide complete sterilization of water at a retention time of 60 minutes, and a very high percent reduction at a retention time of even 2 seconds. It is, however, costly but a combination of short-time sonation

and ultraviolet light may be considered. Chemical disinfection can be attained by using a variety of chemicals.

Oxidizing Chemicals. Oxidizing chemicals are:

- the halogens—chlorine, bromine and iodine
- ozone
- other oxidants such as potassium permanganate and hydrogen peroxide

Gaseous chlorine and several chlorine compounds are the most economical and useful. Chlorine dioxide (ClO_2) produces no taste and odor but has to be used instantaneously since it is unstable. ClO_2 is superior to an equivalent concentration of free Cl_2 when dealing with alkaline waters (pH < 7.5). Bromine (Br_2) and iodine (I_2) have been used to a very limited degree for the disinfection of swimming pool waters. Iodine tablets are used on individual levels. Ozone is acceptable but relatively expensive and requires instant use. An advantage is that it has no taste and odor problems. Potassium permanganate is relatively less efficient, more expensive, and imparts taste, odor, and color to water. Hydrogen peroxide is a strong oxidant but a poor disinfectant and is relatively expensive.

Metal Ions. Silver ions are bactericidal but silver is very costly or ions are strongly algicidal but only weakly bactericidal. However, copper is relatively cheaper and also toxic. Both compounds have some human toxicity.

Alkalis and Acids. Pathogenic bacteria do not survive for long in highly alkaline (pH > 11) or highly acidic (pH < 3) waters. There is incidental destruction of bacteria in the lime softening process which is an example.

Surface-Active Chemicals. Detergents are surface-active chemicals. The cationic detergents are strongly destructive; neutral detergents intermediate; the anionic detergents are weak in their action. Detergents are used only in the wash waters and the rinse waters of food establishments, dairies, soft drink establishments, and so on.

Among the disinfection methods described, chlorination is the universally accepted method of disinfection.

Chlorination

Water disinfection by chlorination was first used in 1908 for Jersey City in the United States. Chlorine is a greenish yellow toxic gas found in nature only in the combined state, chiefly with sodium as common salt. It has a penetrating and irritating odor, is heavier than air, and can be compressed to form a clear, amber-colored liquid. As a dry gas chlorine is noncorrosive but in the presence of moisture, it becomes highly corrosive. Chlorine is slightly soluble in water. The effect of chlorine on the pH of water is shown in Figure 5-24.

Free Residual Chlorine. Chlorine reacts with water to form hypochlorous acid (HOCl) and hydrochloric acid (HCl) according to the following reactions:

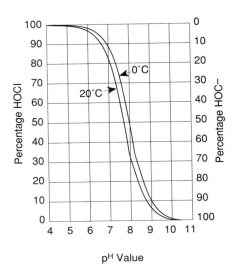

Figure 5-24 Effect of pH on chlorine-water reactions

$$Cl_2 + H_2O = HOCl + HCl$$
$$HCl = H^+ + Cl^-$$

The hypochlorous acid dissociates into hydrogen ions (H^+) and hypochlorite ions (OCl-) according to

$$HOCl = H^+ + OCl^-$$

Pathogenic organisms in water that may result in problems are summarized in Table 5-9.

Undissociated HOCl is about 80-100 times effective as a disinfectant than OCl ions. At a pH value of 5.5 and below, it is practically 100 percent unionized HOCl while above pH 9.5, it is all OCl ions. Between pH 6.0 to 8.5, a very sharp change occurs. The addition of chlorine does not result in any significant change in the pH of the natural waters because of their buffering capacity. Free available chlorine may be defined as the chlorine existing in water as hypochlorous acid and hypochlorite ions.

Free chlorine reacts with ammonia, proteins, amino acids, and phenol, if present in water, to form chloramines and chloroderivatives and provides combined available chlorine. Reactions with ammonia include:

$$Cl_2 + H_2O = HOCl + HCL$$
$$NH_3 + HOCl = NH_2Cl + H_2O$$
$$NH_2Cl + HOCl = NHCl_2 + H_2O$$
$$NHCL_2 + HOCl = NCl_3 + H_2O$$

Table 5-9 Pathogenic Organisms in Water

Microorganisms	Resulting Problems
Actinomycetes	Undesirable tastes and odors.
Algae	Undesirable taste and odors, clog filters.
Coliform bacteria	Indicator organism grouping used as an index for the hygienic quality of water. A high coliform count is presumptive evidence of fecal contamination.
Fecal streptococci	An indicator of fecal contamination; sometimes used as an alternative to coliform index.
Iron bacteria	Produces slimy, often red-colored growth in wells and water mains. Levels of 0.1-0.2 mg/1 iron are sufficient for growth.
Vibrio chlorea	Cholera; initial waves of cholera epidemics are waterborne; secondary cases by contact, food, and flies.
Salmonella typhi	Typhoid fever. Principle modes of transmission are food and water.
Shigella dysenteriae, Shigella flexneri, Shigella boydi, Shigella sonnei	Bacilary dysentery (shigellosis) Fecal-oral transmission via water, milk, food, flies, or direct contact.
Salmonella paratyphi	Paratyphoid fever. A few outbreaks are waterborne.
Entamoeba histolytica	Causes amebic dysentery. Endemic cases are by personal contact, food, and possibly flies. Rare epidemics are mainly waterborne.
Virus not isolated	Infectious hepatitis can be transmitted by water, milk, and food.

Monochloramine (NH_2Cl) and dichloramine ($NHCl_2$) have disinfectant properties. Their effect is considerably less but lasts longer than that of free chlorine. Trichloramine (NCl_3) has no disinfectant properties at all. Trichloramine is found at a pH value below 4.4, dichloramine predominates for pH between 5 and 6.5, and monochloramine predominates for pH over 7.5. The ratio of ammonia to chlorine is 1:4 for optimum results. The minimum concentrations of free and combined residual chlorine for effective disinfection are given in Table 5-10.

Table 5-10 Minimum Concentrations of Free and Combined Residual Chlorine for Effective Disinfection

pH Value	Minimum Concentration of Residual Chlorine (ppm)	
	Free[a]	Combined[b]
6-7	0.2	1.0
7-8	0.2	1.5
8-9	0.4	1.8
9-10	0.8	not recommended
>10	>0.8 with longer contact time	not recommended

[a] Minimum contact time of 10 minutes.
[b] Minimum contact time of 60 minutes.

Some compounds give up chlorine:

- Bleaching powder with pressure gauges indicating the cylinder and delivery pressures feed a combination of lime and chlorine, containing 25 percent to 35 percent chlorine
- High-strength calcium hypochlorite, a relatively stable compound, containing 60 percent to 70 percent chlorine by weight

In plain chlorination, chlorine is applied to water as the only treatment for public health protection. It can be practiced when:

- Turbidity and color of the raw water are low, (turbidity: 5-10 NTU).
- Raw water source is relatively unpolluted.
- Organic matter is less, and iron and manganese do not exceed 0.3 mg/l.

In prechlorination, chlorine is applied to the water prior to any treatment process. The chlorine may be added in the suction pipes of raw water pumps or to the water as it enters the mixing chamber. Prechlorination improves coagulation and reduces taste, odor, algae, and other organisms in all treatment units.

Postchlorination usually refers to the addition of chlorine to the water after all other treatments. It is a standard practice at the rapid sand-filter plants.

Heavy chlorination (superchlorination) is required for quick disinfectant action or the destruction of tastes and odors and may produce high residuals. This process is termed *super-chlorination*. It is practiced in emergency during a breakdown or when the water is heavily polluted. The dose of chlorine may be as high as 10-15 mg/l with a contact period of 10-30 minutes. The undesirable excess chlorine in the water has to be removed by dechlorination.

Dechlorination. The excess chlorine resulting from superchlorination is removed through dechlorination by reducing agents such as sulphur dioxide, sodium bisulphite ($NaHSO_3$), sodium sulphite (Na_2SO_3), sodium thiosulphate ($Na_2S_2O_3$); oxidizing processes such as activated carbon; and aeration.

Rechlorination. Rechlorination is useful when the distribution system is large and it may be difficult to maintain a residual chlorine of 0.2 mg/l at the farthest end, unless a very high dosage is applied at the postchlorination stage. This is costly and may make the water unpalatable, at least in the earlier parts of the distribution system. Rechlorination is carried out in the distribution system at service reservoirs, booster pumping stations, or any other convenient points.

The most common practice in public water supplies is the addition of chlorine either in gaseous form or in the form of a solution made by dissolving gaseous chlorine in a small auxiliary flow of water, the chlorine being obtained from cylinders containing the gas under pressure. The normal chlorine dosage required to disinfect water not subject to significant pollution would not exceed 2 mg/l. The actual chlorine dose requirement is obtained from the chlorine demand tests.

Chlorinators. A chlorinator is a device designed for feeding chlorine to water. It regulates the flow of gas from the chlorine container. Fittings and parts of a chlorinator consist of:

- Chlorine cylinders with 45-1,000 kg capacity or tank cars with 15,000, 30,000, or 50,000 kg capacity.
- Fusible plug, a safety device, provided to the cylinders designed to melt or soften between 70°C to 75°C to avoid the pressure build-up.
- Reducing valve to bring down the gas pressure to 700-300 N/mm^2.
- Pressure gauges indicating the cylinder and delivery pressures.
- Orifice meter or rotameter to measure the flow rate.
- A desiccator valve or a nonreturn valve containing concentrated sulphuric acid or calcium chloride through which the chlorine must pass to free it from moisture to prevent corrosive action on the fittings.

Other parts of the system will consist of:

- Piping system—As the chlorine gas is very corrosive in the presence of moisture, the piping is of a suitable PVC material.

- Chlorine vaporizers—When the chlorine requirement exceeds 900 kg/d, or when the rate of gasification is less than required because of low room temperature, chlorine vaporizers such as thermostatically controlled hot water baths are provided.
- Chlorine feeders—Chlorine feeders are used to control or measure the flow of chlorine in the gaseous state. Direct feed equipment and solution feed equipment are the common types.
- Chlorine housing—The chlorine cylinders and feeders should be housed in a fire resistant building that is easily accessible, close to the point of application, and convenient for truck loading and unloading. It should also have proper ventilation.
- Safety—The necessary safety precautions should be observed during the chlorination process.

FLUORIDATION

The addition of fluoride chemicals to public water supplies to prevent dental caries is an established public health practice in this and many other countries. Very few health measures have induced so much controversy, active support, and opposition as fluoridation. Not even the first use of chlorine as a sterilization agent in water early in the twentieth century produced as much discussion.

A fluoride ion naturally or artificially present in drinking water is absorbed to some degree in the bony structure of the body including the tooth enamel. The absorption of fluoride by the teeth is most efficient during the formative stage although there is some absorption at all ages. The beneficial effect of the fluoridation of water is most pronounced when the children have consumed the fluorine water from infancy.

A number of compounds are available which, when they are dissolved in water, release fluoride ion (F-). Five of these are commonly used for treating water: sodium fluoride (NaF), sodium silicofluoride (Na_2SiF_6), ammonium silicofluoride ($NH_{42}SiF_6$), fluosilicic acid (H_2SIF_6), and calcium fluoride ($CaFe_2$). Sodium silicofluoride and ammonium silicofluoride are sometimes called sodium fluosilicate and ammonium fluosilicate, respectively. Fluosilicic acid is also called hydrofluosilicic acid. All of these chemicals except fluosilicic acid are normally supplied in bags or drums in powder or crystalline form. Fluosilicic acid can be supplied in tank cars, tank wagons, or drums as a water solution (usually 25 percent to 30 percent concentration). These chemicals have many industrial uses but grades suitable for some industrial purposes may be quite unsatisfactory for water treatment. It is of importance that they be purchased under specifications established by the American Water Works Association to ensure satisfactory application.

Chemicals received in dry form are usually stored in their original containers until they are used. They are then transferred to bins over dry-feed machines. It is highly important that workers engaged in filling the bins be well protected from dust during filling operations. Face masks to prevent dust inhalation, tight clothing to protect skin surfaces, and rubber gloves should be worn by these workers. In larger plants dust exhaust systems are essential since the amount of the chemicals transferred on a given occasion is often quite large.

Chemicals received in solution form are pumped into storage facilities and no special problems are encountered provided the equipment used is satisfactory for the purpose.

Fluoride salts in solution and fluosilicic acid are very corrosive to metals. Plastic containers and piping are advisable for all concentrated solutions.

There are three general methods of applying fluoride chemicals:

- In dry form as received. This method is popular especially for sodium silicofluoride, although it is also used extensively for sodium fluoride. It requires well designed accurate equipment which may be manually controlled if the flow is constant. More elaborate automatic dry-feed equipment is needed if the flow is variable.
- In solution form. In this system the chemical is in water solution either as received or made in the plant. Feeding is easier, since relatively simple metering pumps are used. Careful control is necessary, however, to insure accurate dosing. Sodium silicofluoride and calcium fluoride are rarely applied in this manner because they are only slightly soluble. It is the usual method, however, for the application of fluosilicate acid and also is commonly used for sodium fluoride.
- As a slurry. It is sometimes advantageous to dispense the chemicals in the form of a slurry, which might be described as a water suspension of a powder. For uniform dosing, continuous stirring of the slurry mixture and accurate dispensing are required.

The U.S. Public Health Service recommends that where fluoridation is practiced, the average fluoride concentration must be kept within the upper and lower control limits as shown in Table 5-11.

A reduction of 60 percent to 65 percent in dental cavities can be expected from fluoridation. However, the presence of fluoride in average concentrations greater than two times the optimum values in Table 5-11 should constitute grounds for rejection of the water supply.

Setting equipment feed for the calculated dosage is not in itself sufficient control for fluoridation. The fluoride ion concentration of the water after thorough mixing must be determined by chemical testing at least once daily. Systems which require manual control of feeding also require a chemical test of residual fluoride after each dosage change. This test should be carried out by the plant chemist or plant operators, and commercial testing equipment is available for the purpose.

Once a routine has been established for the control of dosing and for the measurement of the fluoride-ion concentration, most problems associated with fluoridation involve either corrosion of feedline piping and equipment or buildup of insoluble deposits in the feeding equipment. Calcium fluoride deposits are usually encountered when a hard water is used to make up a concentrated fluoride solution for dosing the main stream in the plant. A convenient remedy is to pass the make-up water through an ion-exchange softener to remove calcium before adding fluoride salts.

Fluosilicic acid is sometimes supplied with an excess of silicic acid (SiO_2) which precipitates when the material is diluted with water. If the acid is injected into the plant stream without dilution, no precipitation occurs. In small plants, it is necessary to dilute the strong solution to permit an accurate control of the

Table 5-11 Recommended Control Limits for Fluorine

Annual average of maximum daily air temperatures*	Recommended control limits fluoride concentrations mg/l		
	Lower	Optimum	Upper
50.0-53.7	0.9	1.2	1.7
53.8-58.3	0.8	1.1	1.5
58.4-63.8	0.8	1.0	1.3
63.9-70.6	0.7	0.9	1.2
70.7-79.2	0.7	0.8	1.0
79.3-90.5	0.6	0.7	0.8

[a]Based on temperature data for a minimum of five years.

dosage and the resulting precipitate clogs the feeding equipment at various points. The use of softened water or even distilled water for dilution has no beneficial effect; precipitation still occurs. The remedy is to add two parts by volume of hydrofluoric acid to 100 parts of 30 percent fluosilicic acid. Hydrofluoric acid is a highly corrosive chemical and it is dangerous for persons to use without special training. It is much better to purchase fluosilicic acid which has already been treated with hydrofluoric acid by the supplier.

Corrosion of feedline piping and metal equipment by solutions of various salts, including fluoride salts, is a common phenomenon. Corrosion-resistant metals may be substituted for common metals or plastic may be used in place of metal. A plastic or rubber coating over base metal is also effective.

ION EXCHANGE—DEMINERALIZATION

The reactions involved in the ion-exchange process are

$$Ca^{++} + 2(Na^+R^-) \rightarrow (Ca^{++} R_2^{--}) + 2 Na$$
$$Mg^{++} + 2(Na^+R^-) \rightarrow (Mg^{++} R_2^{--}) + 2 Na$$
$$OH^- + (R^+CL^-) \rightarrow (R^+OH^-) + Cl^-$$
$$SO_4^{--} + 2(R^+OH^-) \rightarrow (R_2^- SO_4^-) + 2 OH^-$$

In these reactions, the letter R is used to indicate the anionic or cationic portion of the material which does not enter into the reactions. Thus, NaR means the base-exchange material containing sodium, whereas CaR_2 and MgR_2 represent the same material after the sodium has been exchanged for calcium and magnesium by the removal of these minerals from water and the release of an equivalent amount of sodium to water. To restore the sodium, regeneration is practiced in which solutions of common salts are used.

The advantages of the ion-exchange process are simplicity of operation and control, compactness in size, and ability to demineralize waters to the desired extent. Noncarbonate hardness is more economically removed by the ion-exchange process than with soda ash. However, the process is not economical when total hardness exceeds 850-1,000 ppm. The total solids content should be below 3,000-5,000 ppm. The applied water should not contain turbidity more than 5 NTU. Green sand and zeolites should not be used to soften waters of high sodium alkalinity or those with a pH below 6.0 or even 8.0. Residual chlorine reacts with some of the organic exchange materials and therefore chlorine should not be applied to raw waters when treated by such materials.

Alumina, SiO_2, MnO_2, metal phosphates and sulphides, lignin, proteins, cellulose, wool, living cells, carbon, and resins have ion-exchange capacity. Synthetic zeolites are reasonably stable and have high capacities; hence, they are commercially used.

Exchangers are of two types:

- Cation exchangers
- Anion exchangers

Cation exchangers contain negatively charged functional groups, such as SO_3 from H_2SO_4, or carboxylic group derived from weak acid. Anion exchangers contain positively charged functional groups, such as quaternary ammonium (NR_3^+), imino (NRH_2^+), phosphonium (PR_3^+), and sulphonium (SR_3^+).

Demineralization is effected in a two-step process in which water is passed successively through a cation exchanger in the H^+ form (H^+R^-); and an anion exchanger in the OH^- form (R^+OH^-).

Mixed-bed exchanger is where a single column contains a mixture of equivalent quantities of cation and anion exchangers. The effluent is generally superior in quality. To regenerate a mixed bed, the resins must be separated. This can be done by differential backwashing because cation-exchange and anion-exchange resins have different densities. The cation-exchange resins are regenerated with a strong acid, for example, H_2SO_4. If weak acids such as CO_2 and silicic acid

$Si(OH)_4$ are to be removed, strongly basic anion exchangers must be employed. These must be regenerated with sodium hydroxide. Some weakly basic anion exchangers can be regenerated with soda ash.

6 GROUNDWATER

Groundwater contamination is so severe in certain localities that continued use of such water can lead to serious health problems. Even though serious groundwater pollution problems may exist over relatively small geographic areas, they often occur in areas having the highest population densities. Figure 6-1 shows the populations served by groundwater for domestic supply.

WITHDRAWAL AND CONSUMPTION

The United States relies on both surface and groundwater for domestic use, industry, and agriculture. Fresh-water withdrawals totaled approximately 338,000 million gallons per day in 1985 for a population of just over 242 million persons in the United States or about 1,400 gallons per capita per day. This per capita is deceptively high because it includes water use for domestic fresh-water withdrawals. A more accurate consumptive measure is 78 gallons per day and 105 gallons per day for self-supplied and publicly supplied water, respectively. Figure 6-2 shows national water withdrawals from 1960-1985 for both surface and groundwater.

Groundwater is a valuable resource that is withdrawn in all parts of the country for a variety of uses. Groundwater is specially important in a selected number of locations for a specific number of water users, including self-supplied domestic, commercial, and livestock users in particular. As this resource becomes more contaminated and more scarce, demand for high-quality water will continue to grow making groundwater even more valuable and protection more important.

GROUNDWATER HYDROLOGY

A common misconception is that groundwater takes the from of underground lakes, streams, and rivers flowing through tunnels and caverns. Groundwater can be defined simply as the water under the surface of the earth that freely fills porous spaces within the ground. Groundwater occurs because earthen material is not entirely solid; rather, it is partially composed of void space as a result of its granular texture or fractures.

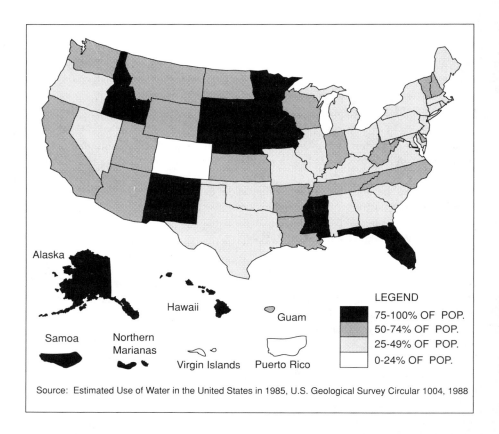

Figure 6-1 Percentage of state and territory populations served by groundwater for domestic supply

Figure 6-3 schematically shows water passing through the top layer of material under the surface of the front because of capillary and gravitational forces and is not held there. This area is called the *vadose zone* or unsaturated zone. Below the point referred to as the water table, however, all the void space within the material is saturated with water. This area is referred to as the *saturated zone*. It is the water within the saturated zone only that is considered to be groundwater. Water moving through the unsaturated zone and found in the *capillary fringe* just above the water table is not included within the definition of groundwater. An analogy is that of a partially wet sponge, where some of the void space contains gases, that is, the unsaturated zone and the rest is saturated with water.

Groundwater always flows from areas of relatively high pressure (*upgradient*) to areas of lower pressure (*downgradient*). This hydraulic pressure is caused by

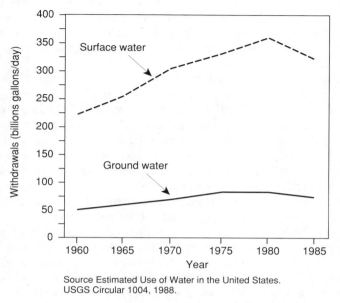

Source Estimated Use of Water in the United States. USGS Circular 1004, 1988.

Figure 6-2 National water withdrawals (1960-1985)

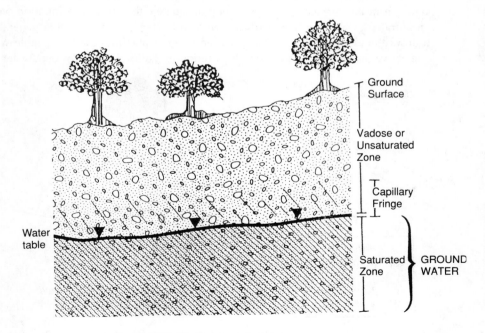

Figure 6-3 Underground water zones

gravitational forces acting upon the water as it moves through porous materials and around impermeable barriers. Groundwater moves through a matrix of earthen materials and is not free flowing as is surface water. Typical velocity at which groundwater flows is relatively slow, ranging from a few feet to fractions of an inch per day.

A common measure of groundwater velocity, especially with regard to the movement of groundwater toward a well, is referred to as *time of travel*. Environmental planners and hydrologists are concerned with the length of time it will take a groundwater contaminant to reach a well. Researchers often specify the time of travel of concern and then determine the distance from the well that this represents. For example, the six-month, one-year, and five-year time of travel points for a given well might be from zero to miles upgradient of the wellhead. Because groundwater velocity is highly dependent on site-specific hydrogeology, and to a lesser extent precipitation and pumping rates, the relevant distances for a fixed time of travel will be highly variable from well to well.

Groundwater bodies generally are not distinct or isolated units of water but rather are part of the hydrologic cycle. The hydrologic cycle describes the movement of water in its various forms (for example, rain, snow, vapor) through the environment. Figure 6-4 schematically illustrates how water generally falls to the earth as precipitation, flows along the ground as surface water or infiltrates into the earth as groundwater, and ultimately evaporates back into the atmosphere to complete the cycle. Groundwater is *recharged* primarily by the infiltration of water from the surface. This water may originate either from precipitation onto unsaturated ground or from surface bodies of water. The water table generally follows the topography of the surface; however, sometimes the water table will rise above the surface of the ground, resulting in the *discharge* of groundwater at

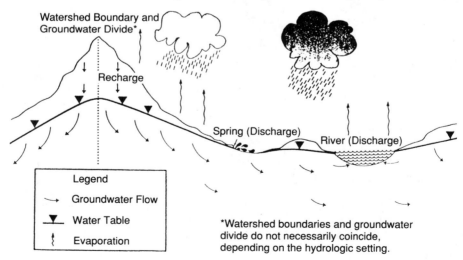

Figure 6-4 Recharge and discharge of groundwater

a seep or spring or into an existing surface water body (such as a lake or river). Figure 6-4 also illustrates how these types of discharge and recharge occur. Because precipitation tends to fluctuate over the seasons, the water table also will fluctuate over time, and points of recharge and discharge may change. Seeps may dry up during a drought, and groundwater may both discharge to or be recharged by a surface water body depending on the height of the water table. Generally, shallow groundwater will discharge to a surface water body during periods of drought and be recharged by that same surface water body during periods of high precipitation.

A point of reference for studying and understanding surface water flow and recharge is the watershed. A watershed can be defined as the land area draining into a single surface water body. An example of a small watershed is that of a valley draining into a single river. Depending on the area of study, larger watersheds can be defined as a lake. Because groundwater generally is recharged by infiltration from the surface, shallow groundwater flow typically follows surface water drainage patterns, that is, shallow groundwater flows in the same direction as the surface water above. Moreover, groundwater *divides* typically coincide with watershed boundaries. Precipitation falling on the top of a mountain ridge will tend to move either side of that distinct watersheds; surface water drainage, shallow bodies of groundwater can recharge deeper, regional groundwater bodies. Regional water bodies, however, may flow in a different direction from more shallow water, as was shown in Figure 6-4 dependent on the hydrogeologic setting of the site.

Groundwater flow and recharge are important for protecting a source of groundwater withdrawal, such as a naturally occurring spring or a man-made well, or when seeking to determine how best to intercept spills and leaks of contaminants. Flow and recharge are important because they will dictate where sources of contamination to the withdrawal point might originate. The activity of withdrawing water, particularly from a pumping well, can affect flow and recharge patterns. Hence, once a well is sunk and a pumping rate is established, the recharge area for that well can be identified. In essence, the recharge area will consist of all the land area (and/or surface waters) from which precipitation infiltrating through the ground will eventually reach the well. Recharge areas are predominantly upgradient of the well, that is, above the well with regard to the groundwater flow pattern. Depending on the hydrogeologic setting and pumping rate, some amount of water downgradient of the well also will be drawn back to the well as a result of pumping.

The term *recharge* generally refers to replenishing an entire groundwater body; hydrogeologists usually use the term *zone of contribution* to describe the area recharging a specific well or spring. The zone of contribution can be mapped both from an aerial view and in cross section, as illustrated in Figure 6-5. Shown here is how the activity of pumping can change the contour of the water table by creating a *cone of depression* and the direction of groundwater flow immediately surrounding the well itself. Planners using empirical data to establish *wellhead*

172   *Water Management and Supply*

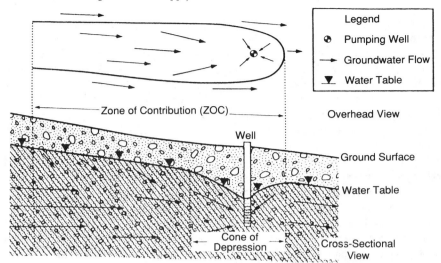

Figure 6-5 Overhead cross-section view of groundwater flow at a pumping well

protection areas typically use the zone of contribution because this zone by definition delineates the area within which introduced pollutants of contamination will eventually reach the well. The problem with using this method for delineating a wellhead protection area is that it requires extensive sampling and hydrogeologic analysis.

Aquifers

While bodies of groundwater may be found in many kinds of geologic formations, not all saturated formations are considered to be *aquifers*. An aquifer can be defined as a geologic formation that can store, transmit, and yield significant quantities of water to a well or spring. Significant generally means a usable supply of water, so a formation yielding as little as the amount of water necessary to support a single family with 100 gallons per day might be considered an aquifer, although water supply aquifers often yield several millions of gallons per day. Geologic formations such as layers of consolidated sedimentary rock, unfractured bedrock, or clay deposits, which do not store or yield significant amounts of water and impede the flow of water, are commonly referred to as *aquitards*. The ability of an aquifer to store and transmit water is primarily a function of its *porosity* and *permeability*.

Porosity describes the proportion of soil or material that is void space where water can collect. For example, an unconsolidated formation such as sand and gravel might be 60 percent solid material and 40 percent pore space. In contrast, formations with low porosity such as dense crystalline rock might consist of only 5 percent void space.

Permeability is the ability of the medium to transmit water. The permeability of an unconsolidated medium is a function of the size and shape of the rock particles, the degree of uniformity in the size and shape of the particles, and the pattern in which the particles are packed. Unconsolidated media such as sand and gravel stratified drift, along with fractured and porous bedrock, typically comprise the aquifers used for water supplies because of their high degree of porosity and permeability.

Aquifers are often separated, for purposes of comparison, into two general classes: unconfined and confined aquifers. An unconfined aquifer is characterized by the absence of a flow-restricting material such as an aquitard above it and a free water table that will rise and fall subject to atmospheric influences (after a rainfall).

Figure 6-6 illustrates an unconfined aquifer, where the water level in a well sunk in an unconfined aquifer will rest at the water table. A confined aquifer, also referred to as an *artesian* aquifer, is defined as an aquifer between two horizontal confining layers such that no free surface exists.

Figure 6-7 illustrates a confined aquifer, that if a well is sunk is not a confined aquifer and the surface of the ground is below the *piezometric surface* where the water table would be were it not for the confining layer. The water will flow freely out of the well. Such wells are commonly referred to as artesian wells. If the surface of the ground is above the piezometric surface, then the water level in the well will rise only to the piezometric surface.

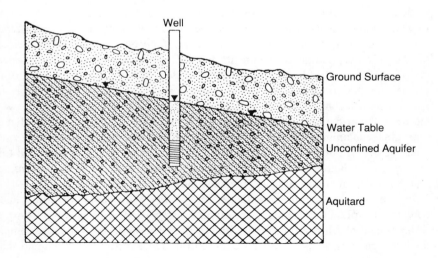

Figure 6-6 Unconfined aquifer

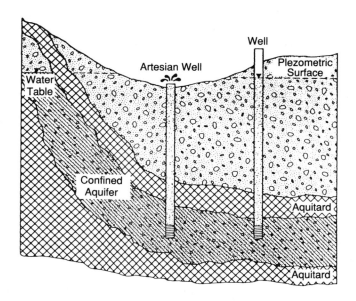

Figure 6-7 Confined aquifer

CONTAMINATION

Groundwater contamination can occur by intentional or unintentional introduction of contaminants directly into the groundwater through an active or abandoned well, by the infiltration of contaminants through the unsaturated zone, or by the flow of contaminants from interconnected surface waters. Once a contaminant enters the ground, it tends to move in a distinct *plume* because of the slow velocity of groundwater, unlike contaminants in surface water which tend to disperse quickly. The extent of the resulting contamination, both in terms of the size of the plume and the concentrations of the contaminants, will be a function of site-specific hydrogeologic factors such as matrix consolidation, depth to the water table and the properties of the contaminants themselves, and the processes that the individual contaminants undergo while in both the saturated and unsaturated zones. Such factors will determine whether a contaminant remains in a concentrated form or disperses; whether it stays at the top of the aquifer or sinks to the bottom, and whether it retains its identity or degrades into other products.

Figure 6-8 shows four potential effects of the hydrogeologic setting and a contaminant plume as it might appear in cross section as it moves toward the water table through stratified soils with varying permeabilities. The more complex the hydrogeology at a site, the more difficult it is to estimate the extent and shape of a contaminant plume.

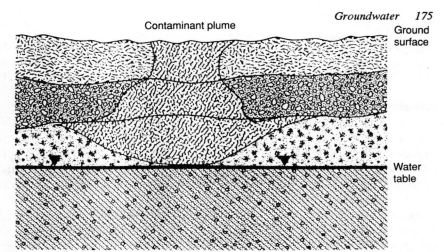

Figure 6-8 **Contaminant seepage through stratified soils with varying permeabilities**

In addition to site-specific hydrogeology, the unique properties of the contaminant itself will affect how it moves through both unsaturated soils and groundwater. The properties determining a contaminant's mobility and attenuation include:

- volatility
- viscosity
- solubility
- density
- ionic size and charge
- chemical reactivity
- radioactive decay rate
- biodegradability

As contaminants pass through unsaturated materials and into a saturated zone, a variety of biotic processes (for example, bacterial degradation) and abiotic processes (for example, absorption) take place. Depending on the contaminants properties, these processes will determine the extent and nature of contamination. Highly volatile contaminants may volatilize entirely into the air before reaching the saturated zone, while highly viscous contaminants tend to move slowly through porous material. Contaminants less viscous than water such as gasoline move readily through unsaturated material. Highly soluble contaminants dissolve and disperse quickly throughout an aquifer; dense contaminants tend to sink through the saturated zone. Depending on ionic size and charge, some contaminants tend to bind by physical absorption or chemical absorption to particles, thus attenuating the contamination of water. Other contaminants may react chemically and precipitate, neutralize or degrade radioactively. Some contaminants may be degraded biologically by organisms (for example, bacteria) found in either the saturated or unsaturated zones, although the by-products of such degradation may in fact be more toxic than the original contaminants.

Sources of Contamination

As water moves through the hydrologic cycle, its quality changes in response to differences in the environments through which it passes. The changes may be either natural or man-made. While in some cases they can be controlled, in other cases they cannot, but in most instances they must be managed in order to limit adverse changes water quality.

The physical, chemical, and biological quality of water may range over wide values. It is often difficult if not impossible to distinguish the origin of many water quality problems. *Natural* quality reflects the types and amounts of soluble and insoluble substances with which the water has come into contact. Surface water in general contains less dissolved solids than groundwater, although at certain times where groundwater runoff is the major source of stream flow, the quality of both surface water and groundwater is similar.

During periods of runoff, streams may contain large quantities of suspended materials and under some circumstances a large amount of dissolved solids. Usually, however, during high rates of flow, streams have a low dissolved mineral concentration.

Although the chemical quality of water at the surface or shallow aquifers may range within fairly broad limits from one time to the next, deeper groundwater may be characterized by nearly constant chemical and physical properties at least on a local scale where the aquifer is unstressed by pumping.

The quality of groundwater can be significantly affected by waste disposal and land use. One major source of contamination is the storage of waste materials in excavations such as pits or mines or from leaking underground storage tanks. Water soluble substances that are dumped, spilled, spread, or stored on the land surface may eventually infiltrate. Groundwater also can become contaminated by the disposal of the fluids through wells and in limestone terrains through sinkholes directly into aquifers.

A further source of groundwater deterioration is pumping, which may result in the migration of more mineralized water from surrounding strata to the well. Groundwater quality problems that come from the land surface include:

- infiltration of contaminated surface water
- land disposal of solid and liquid waste materials
- stockpiles, tailings, and spoil
- dumps
- disposal of sewage and water treatment plant sludge
- salt spreading on roads
- animal feedlots

- fertilizers and pesticides
- accidental spills
- particulate matter from airborne sources

Groundwater quality problems that originate above the water table include:

- septic tanks, cesspools, and privies
- surface impoundments
- landfills
- waste disposal in excavations
- leakage from underground storage tanks
- leakage from underground pipelines
- artificial recharge
- sumps and dry wells
- graveyards

Groundwater quality problems that originate below the water table include:

- waste disposal in wet excavations
- agricultural drainage wells and canals
- well disposal of wastes
- underground storage
- secondary recovery
- mines
- exploratory wells and test holes
- abandoned wells
- water supply wells
- groundwater development

Infiltration of Surface Water

The yield of many wells tapping stream-side aquifers is sustained by the infiltration of surface water. More than half of the well yield may be derived directly by induced recharge from an adjacent water source which may be contaminated. As the induced water migrates through the subsurface, few substances may be diluted or removed. Filtration is not likely to occur if the water table flow is through openings such as those in some carbonate aquifers. Chloride, nitrate, and several organic compounds which are highly mobile move freely with the water and are not removed by filtration.

A cause of groundwater contamination can be the disposal of waste materials directly onto the land surface. Examples include manure, sludges, garbage, and industrial wastes; waste may occur as individual mounds or it may be spread over the land. If the waste material contains soluble substances, they may infiltrate through the soil.

An example of groundwater contamination caused by stockpiled materials or wastes is the unprotected storage of deicing salt commonly mixed with sand at maintenance lots. The salt readily dissolves as infiltrate or runoff. Stockpiled salt may contain 100 to 300 tons of salt with anticaking additives such as ferric ferrocyanide and sodium ferrocyanide with perhaps phosphate and chromate to reduce corrosivity.

Environmental effects of landfills have become a concern. As rainwater infiltrates through garbage in dumps, it accumulates an ample assortment of chemical and biological substances. The resulting leachate may be highly mineralized and as it infiltrates some of the substances it contains may not be removed or degraded.

Sludge is the residue of chemical, biological, and physical treatment of municipal and industrial wastewaters. They include lime-rich material from water treatment plants as well as sludge from wastewater treatment plants. Sludge typically contains partly decomposed organic matter, inorganic salts, heavy metals, bacteria, perhaps viruses, and nutrients. Nitrogen in a municipal sludge may vary from 1 percent to 7 percent. Land application of wastewater and sewage sludge is an alternative to conventional treatment and disposal and is in common practice by the canning and vegetable industry, petroleum refining, pulp and paper, and the power industry. Infiltration from wastewater stabilization ponds can also be a source of groundwater contamination.

The extent and nature of contamination can be very site-specific and dependent on the types of contaminants released. The greater the variety of geologic formations at a site, the more difficult it will be to characterize the contamination and predict the movement of contaminant plumes.

THE SAFE DRINKING WATER ACT (SDWA)

The SDWA is a line of defense in efforts to ensure that human health is not adversely affected by waterborne pollutants. The regulation is based on maximum contaminant levels (MCLs) and maximum contaminant level goals (MCLGs), which are specific numeric standards for constituents that have a likelihood of being found in drinking water. MCLGs are to be set at levels that present absolutely no risk to human health; the MCLs are to be set as close to these MCLGs as possible, considering economic and technological feasibility. The law sets the water tap as the point of compliance, but for practical purposes, most of the standards are met through treatment at a centralized treatment facility. The SDWA addresses mainly community water systems, which are those with 25 or more hookups.

The MCLs also serve an important purpose in groundwater protection efforts. Its policy is to use MCLs as reference points in governing activities that may have an impact on potential or actual underground sources of drinking water. This means that decisions regarding the levels of protection or cleanup of groundwater are to be set such that MCLs are attained (for remediation activities) or not exceeded (for protection activities). Variations from this reference point, either in favor of more or less stringent standards, are acceptable as long as such variation is supported by the particular use, value, and vulnerability characteristics of potentially affected groundwater.

SOLE SOURCE AQUIFERS

Section 1424(e) of the SDWA establishes the Sole Source Aquifer Program (SSA) which focuses on those aquifers that are the sole or principal sources of drinking water. The phrase *sole or principal* means that at least 50 percent of the drinking water of an affected population comes from the aquifer in question. Figure 6-9 shows principal sole source aquifers in the United States.

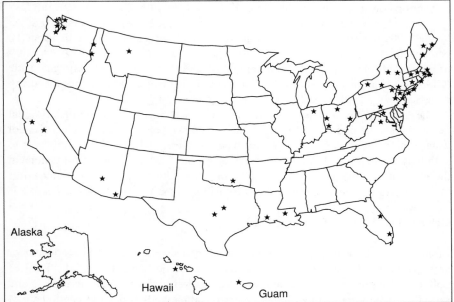

Figure 6-9 Sole source aquifers in the U.S.

The SDWA Amendments of 1986 (Section 1428) introduced a unique, new groundwater protection tool. Dubbed the Wellhead Protection (WHP) Program because of its focus on the areas directly around drinking water wells (wellheads), this tool is unique because it is the first in the federal government's array of congressionally mandated groundwater protection tools to focus comprehensively on the groundwater resource.

Although other programs have been in existence for a longer time, their focus is on specific sources of contamination such as landfills, pesticide application, or underground storage tanks. Even if all these currently regulated sources of groundwater contamination were appropriately controlled, the groundwater might not yet be adequately protected, because of myriad uncontrolled contamination sources.

The SDWA also mandates control of underground injection wells. These wells are used for a variety of purposes, including disposal of hazardous waste, enhancement of oil and gas recovery, and disposal of storm water runoff. The EPA groups injection wells into five categories.

TSCA is considered by many to be the "catchall" statute. This is because any particular chemical or toxic substance with potential for noxious impacts on any environmental media—including, of course, groundwater—can be regulated under the provisions of TSCA. These provisions require manufacturers to supply risk, fate and transport, leachability, and other pertinent data about their chemicals. If the dangers of a chemical are perceived to be too great, the EPA can place restrictions on its use or distribution or can even ban the chemical outright.

The EPA currently regulates approximately 65,000 chemicals under TSCA. Since 1979, the EPA has reviewed about 15,000 new chemicals, with about 10 percent of these resulting in specific regulatory action. The EPA recently promulgated a new rule—Comprehensive Information Assessment Rule (CIAR)—that allows streamlined collection of information about chemicals posing a particularly serious threat.

The Clean Water Act is primarily a surface water protection statute. Its goal demands that the nation's waters be fishable and skippable, two characteristics not usually associated with groundwater. The EPA has had to link groundwater to authority over navigable waters. Nevertheless, there are at least two provisions in the act that are integral parts of the overall groundwater protection framework: Section 106 and the Nonpoint Source Provisions of the Water Quality Act of 1987.

GROUNDWATER MONITORING

Emphasis has been placed on the mitigation of hazards into the groundwater or any usable water supplies. It is important for the determination of the level of pollutant and this is accomplished, in the case of groundwater, by monitoring and sampling. Sampling methodologies, withdrawal methods and equipment, chain of custody, and sample preservation are discussed.

The need for groundwater monitoring can be readily seen from the combination of two factors: (1) the reliance on groundwater as a water resource, and (2) the practice of waste disposal and contamination which continues to provide direct and indirect contamination of groundwater.

Why should groundwater be monitored? The answer is that groundwater is a resource used extensively in the United States. Groundwater usage covers many areas: irrigation, industrial, power generation, etc., but of greatest interest is human consumption. Approximately 50 percent of all drinking water originates as groundwater (96 percent of water by rural domestic users, 37 percent of public water supplies). Since groundwater has been shown to be a significant resource any threat to that resource deserves watching/monitoring from the viewpoint of health and safety.

The specific danger to groundwater is posed by leachate production. Leachate is a liquid or combination of liquids produced by the percolation of water through soils. Leachate production is a function of many different factors but the basic aspects are the amount of water percolating through the landfill and the permeability of the landfill and the soil around it. Leachate is a liquid that is characteristic to a specific site. Table 6-1 shows leachate characteristics and domestic wastewater.

Table 6-2 shows water quality standards acceptable for drinking water. Table 6-3 is a comparison of drinking water standards and composition of various liquid wastes. Table 6-4 categorizes the threats leachate poses to life and property. Figure 6-10 illustrates diagrammatically the hydrogeologic routes of contamination by land waste disposal sites. Figure 6-11 shows, the flow of land-disposed contaminants and Figure 6-12 transport processes at waste disposal sites.

Purposes of Monitoring

The establishment of a specific monitoring system quite often depends on the purpose of monitoring. Basic goals of monitoring are to provide a check on the potential or actual contamination of surface or groundwater but the circumstances surrounding the need to monitor may vary from site to site. Two basic types of monitoring situations present themselves: monitoring existing sites and monitoring new sites.

General purposes of monitoring include:

- Monitoring where the control strategy is the protection of the groundwater or surface water.
- Monitoring where the control strategy emphasizes additionally the protection of the water resource.
- Collection of evidential data for enforcement purposes.
- Demonstration of the presence or absence of leachate contamination as may be done in a statewide evaluation of the extent of the problem.
- Collection of prescriptive monitoring data in response to descriptive monitoring to develop effective engineering remedies for contamination problems.
- Performance of scientific investigations in developing and validating design criteria, such as rates of leachate movement, attentuation, etc.
- Purpose of monitoring at a specific site. Identification and/or definition of groundwater pollution if required.
- Purpose of monitoring for remedial action (cleanup) sites: measure the effectiveness of a remedial action.
- Provide a warning of possible remedial action breakdown.

Table 6-1 Characteristics of Leachate and Domestic Wastewaters

Constituent	Range[a] (mg/l)	Range[b] (mg/l)	Range[c] (mg/l)	Leachate Fresh	Leachate Old	Wastewater	Ratio[d]
Chloride (Cl)	34-2,800	100-2,400	600-800	742	197	50	15
Iron (Fe)	0.2-5,500	200-1,700	210-325	500	1.5	0.1	5,000
Manganese (Mn)	.06-1,400	—	75-125	49	—	0.1	490
Zinc (Zn)	0-1,000	1-135	10-30	45	0.16	—	—
Magnesium (Mg)	16.5-15,600	—	160-250	277	81	30	9
Calcium (Ca)	5-4,080	—	900-1,700	2,136	254	50	43
Potassium (K)	2.8-3,770	—	295-310	—	—	—	—
Sodium (Na)	0-7,700	100-3,800	450-500	—	—	—	—
Phosphate (P)	0-154	5-130	—	7.35	4.96	10	0.7
Copper (Cu)	0-9.9	—	0.5	0.5	0.1	—	—
Lead (Pb)	0-5.0	—	1.6	—	—	—	—
Cadmium (Cd)	—	—	0.4	—	—	—	—
Sulfate (SO_4)	1-1,826	25-500	400-650	—	—	—	—
Total N	0-1,416	20-500	—	989	7.51	40	25

[a] Office of Solid Waste Management Programs, Hazardous Waste Management Division. An environmental assessment of potential gas and leachate problems at land disposal sites. USEPA Pub SW-110 of Cincinnati, USEPA, 33p (Open-file report, restricted distribution) (1973).

[b] Stiener, R. C., Fungaroli, A.A., Schoenberger, R. J. and Purdom, P. W., "Criteria for sanitary landfill development," Public Works, 102(3) 77-79 (March 1971)

[c] Gas and leachate from land disposal of municipal solid waste, summary report Cincinnati, USEPA, Muncipal Environmental Research Laboratory (1975)

[d] Brunner, D. R. and Carnes, R. A., "Characteristics of percolate of solid and hazardous waste deposits." Presented at AWWA (American Water Works Association), 94th Annual Conference, Boston, MA (June 1974)

Table 6-1 Characteristics of Leachate and Domestic Wastewaters (Continued)

Constituent	Range[a] (mg/l)	Range[b] (mg/l)	Range[c] (mg/l)	Leachate Fresh	Leachate Old	Wastewater	Ratio[d]
Conductivity (μmhos)	—	—	6,000-9,000	9,200	1,400	700	13
TDS	0-42,276	—	10,000-14,000	12,620	1,144	—	—
TSS	6-2,685	—	100-700	327	266	200	1.6
pH	3.7-8.5	4.0-8.5	5.2-6.4	5.2	7.3	8.0	—
Alk. as CaCO$_3$	0-20850	—	800-4,00	—	—	—	—
Hardness, tot.	0-22,800	200-5,250	3,500-5,000	—	—	—	—
BOD$_3$	9-54,610	—	7,500-10,000	14,950	—	200	75
COD	0-89,520	100-51,000	16,000-22,000	22,650	81	500	45

[a] Office of Solid Waste Management Programs, Hazardous Waste Management Division. An environmental assessment of potential gas and leachate problems at land disposal sites. USEPA Pub SW-110 of Cincinnati, USEPA, 33p (Open-file report, restricted distribution) (1973).

[b] Stiener, R. C., Fungaroli, A.A., Schoenberger, R. J. and Purdom, P. W., "Criteria for sanitary landfill development," Public Works, 102(3) 77-79 (March 1971)

[c] Gas and leachate from land disposal of municipal solid waste, summary report Cincinnati, USEPA, Muncipal Environmental Research Laboratory (1975)

[d] Brunner, D. R. and Carnes, R. A., "Characteristics of percolate of solid and hazardous waste deposits." Presented at AWWA (American Water Works Association), 94th Annual Conference, Boston, MA (June 1974)

Table 6-2 Drinking Water Standards

Constituent	Recommended concentration limit[a] (mg:l)
Inorganic	
Total dissolved solids	500
Chloride (Cl)	250
Sulfate (SO_4^{2-})	250
Nitrate (NO^-_3)	45[b]
Iron (Fe)	0.3
Manganese (Mn)	0.05
Copper (Cu)	1.0
Zinc (Zn)	5.0
Boron (B)	1.0
Hydrogen sulfide (H_2S)	0.05
	Maximum permissible concentration
Arsenic (As)	0.05
Barium (Ba)	1.0
Cadmium (Cd)	0.01
Chromium Cr^{VI})	0.05
Selenium	0.01
Antimony (Sb)	0.01
Lead (Pb)	0.05
Mercury (Hg)	0.002
Silver (Ag)	0.05
Fluoride (F)	1.4-2.4
Organic	
Cyanide	0.05
Endrine	0.0002
Lindane	0.004
Methoxychlor	0.1
Toxaphene	0.005
2, 4-D	0.1
Phenols 2, 4, 5-TP silvex	0.001
Carbon chloroform extract	0.2
Synthetic detergents	0.5
Radionuclides and radioactivity	Maximum permissible activity (pCi l)
Radium 226	5
Strontium 90	10
Plutonium	50,000
Gross beta activity	30
Gross alpha activity	3
Bacteriological	
Total coliform bacteria	1 per 100 ml

Sources: U.S. Environmental Protection Agency, 1975 and World Health Organization, European Standards, 1970.

[a] Recommended concentration limits for these constituents are mainly to provide acceptable esthetic and taste characteristics.

[b] Limit for NO_3 expressed as N is 10 mg/l according to U.S. and Canadian standards; according to WHO European standards, it is 11.3 mg/l as N and 50 mg/l as NO_3.

Table 6-3 U.S. Public Health Service Drinking Water Standards and Composition of Various Liquid Wastes (in Parts Per Million)

Substance	U.S. Public Health Service Standards[a] Group 1[a,b,c]	Group 2[d,e]	Blackwell[f]	Leachate LW5B Du Page[g]	Leachate LW6B Du Page[h]	Influent Sewage[i]	Effluent Sewage[i]	Slaughterhouse wastes[j]	Chemical plant effluent[k]
Alkyl benzene sulfonate	0.5			0.72	0.30				
Arsenic	0.01	0.05	4.31	<0.10	4.6				
Chloride	250		1,697	1,330	135			320	1,070
Copper	1		0.05	<0.05	<0.05	0.450	0.032		2.1
Carbon chloroform extract	0.2								
Cyanide	0.01	0.2	0.024	<0.005	0.02		0.051		
Fluoride		3.4		2	0.31				800
Iron	0.3		5,500	6.3	0.6	2,600	0.938		51
Manganese	0.05		1.66	0.06	0.06				0.48
Nitrate	45		1.70	0.70	1.60				864
Phenols	0.001								
Sulfate	250		680	2	2			370	8,120
Total dissolved solids	500		19,144	6,794	1,198	0.638	0.366	2,690	16,090
Zinc	5			0.13	<0.10				
Barium		1	8.5	0.80	0.30				

[a] U.S. Dept. of Health, Education and Welfare (1962).
[b] Nitrates exceeding 45 ppm dangerous for infants.
[c] Should not be used if more suitable supplies available.
[d] Larger concentrations should be rejected.
[e] Fluoride is temperature dependent.
[f] Typically represents leachate from infiltration.
[g] Leachate from refuse about six years old.
[h] Leachate from refuse about 17 years old.
[i] Data provided by Metropolitian Sanitary District of Greater Chicago.
[j] Data from files of the Illinois Deptartment of Public Health
[k] Rare earth and thorinum production
[l] Twenty-day BOD for leachate. Other values are five-day BOD.

Table 6-3 U.S. Public Health Service Drinking Water Standards and Composition of Various Liquid Wastes (in Parts Per Million) *(Continued)*

	U.S. Public Health Service Standards[a]			Leachate					
Substance	Group 1[a,b,c]	Group 2[d,e]	Blackwell[f]	LW5B Du Page[g]	LW6B Du Page[h]	Influent Sewage[i]	Effluent Sewage[i]	Slaughter-house wastes[j]	Chemical plant effluent[k]
Cadmium	0.01		<0.05	<0.05	0.05	0	0		
Chromium (Cr^{+4})	0.05		0.20	0.15	0.05	0	0		
Lead	0.05		2.7	0.50	0.50	0.138	0.138		
Selenium	0.01		<0.1	<0.10	<0.10				
Silver	0.05			<0.1	<0.1				198
Ammonium			3,255	4,159	1,011	19	16		760
Alkalinity (as $CaCO_3$)			7,830	2,200	540			440	0
Hardness (as $CaCO_3$)			6	1.20	8.90			66	74
Phosphate									0.97
Titanium			2.20	0.10	0.90				6.4
Aluminum			900	810	74				6,190
Sodium			350	18	7	22.4	11		
Hexane solubles			54,610	14,080	225	104	17	3,700	
Biological oxygen demand[l]			39,680	8,000	40	240	70	8,620	
Chemical oxygen demand				6.3	7.0	7.2	7.4	8.1	6.2
pH									

[a] U.S. Dept. of Health, Education and Welfare (1962).
[b] Nitrates exceeding 45 ppm dangerous for infants.
[c] Should not be used if more suitable supplies available.
[d] Larger concentrations should be rejected.
[e] Fluoride is temperature dependent.
[f] Typically represents leachate from infiltration.
[g] Leachate from refuse about six years old.
[h] Leachate from refuse about 17 years old.
[i] Data provided by Metropolitian Sanitary District of Greater Chicago.
[j] Data from files of the Illinois Deptartment of Public Health
[k] Rare earth and thorinum production
[l] Twenty-day BOD for leachate. Other values are five-day BOD.

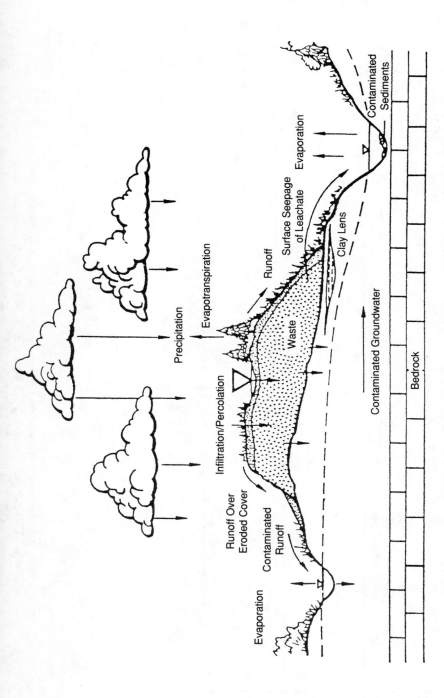

Figure 6-10 Hydrologic pathways for contamination by waste disposal sites

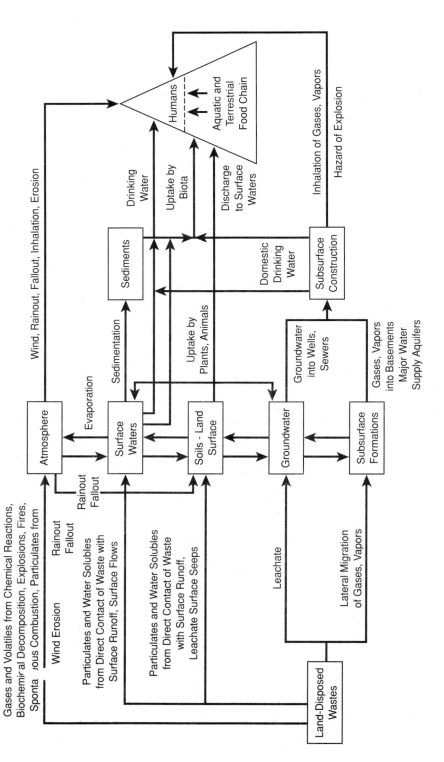

Figure 6-11 Flow of land-disposed waste contaminants through the environment.
Source: U.S.E.P.A., Handbook for Remedial Action at Waste Disposal Sites, Cincinnati, OH: USEPA-1 EPA-625/6-82-006 (June 1982)

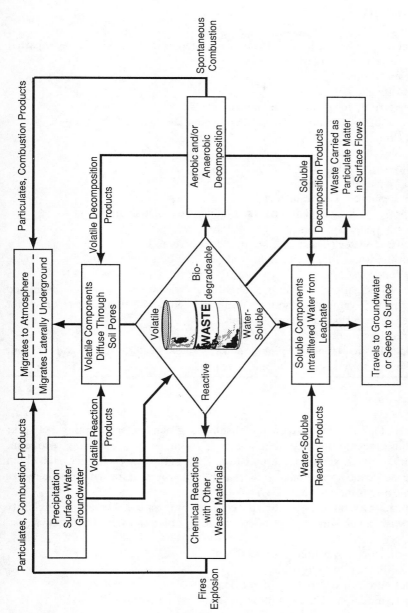

Figure 6-12 Initial transport processes at waste disposal sites. Source: U.S.E.P.A., Handbook for Remedial Action at Waste Disposal Sites, Cincinnati, OH: USEPA-1 EPA-625/6-82-006 (June 1982)

Table 6-4 Potential Damage to Life from Leachate

Damage to Life
  Humans
    acute and chronic health effects, for example, illness, skin damage, partial paralysis, brain damage, death
  Domestic Animals
  Wild Animals
    terrestrial, for example, mammals, birds
    aquatic, for example, fishkills, spawning areas, shellfish, crabs
  Farm Crops
  Other Vegetation
    grasses, shrubs, trees
Physical Damage
  Water Resources Contaminated
    springs, lakes, streams, rivers, groundwater, marshlands
  Drinking Water Supplies
    surface water (reservoirs, springs, lakes, rivers)
    groundwater (domestic, industrial, and public supply wells)
  Land/Water Use
    sport/recreation: fishing, shellfish, swimming, parks decreased property value or facility utilization loss of future use of water resource surface and groundwater
  Material Destruction
    damage to laundry, pipes, sinks, water heater, food, and so on; damage to industrial equipment or products

Prerequisites

Several steps are required in establishing a groundwater sampling network: establishing background, well placement, well design, sampling program, laboratory analysis, and data interpretation. Each of the steps is influenced by both the purpose of monitoring and the research/background information gathered as a prerequisite to monitoring (See Table 6-5).

Establishing a monitoring system is a complex proposition. Therefore, flexibility and professional judgment must be exercised in utilizing each of the four steps:

- initial site inspection
- preliminary investigations
- definition of the hydrogeologic setting
- determination of the pollution potential of the landfill

Table 6-5 Requisites and Uses for Planning Monitoring Wells

Requisite	Use
Probable groundwater flow, direction, and expected seasonal changes	Use this information to areas for background (upgradient) wells and select those where intercepting the leachate plume is likely.
Location of other potential pollutant sources	In conjunction with information on groundwater flow directions, determine if and where pollutants from other sources might be detected in the study area.
Probable groundwater contaminants	Consider the chemical effects of the contaminated groundwater on potential casing materials corrosion, adsorption, and so on, grout, and bentonite seals. Also determine if well design must accommodate special sampling procedures or equipment for these pollutants.
Expected range in depths to groundwater	Use these data to ensure proper well depths for year-round sampling. For water depths below about 20 ft, determine if well design must accommodate submersible pumps or other large down-the-hole equipment.
Types of materials to be drilled	Identify whether consolidated (rock), unconsolidated, or both types of formations will have to be penetrated so that the drilling rig alternatives can be determined. If unconsolidated formations are the only concern, find out if cobbles or boulders are common.
Probable geologic and hydrologic characteristics of the target formation and overlying materials overlying materials	Use these data for selecting screen slot sizes, determining if a gravel pack is necessary, locating where the borehole should be sealed with grout or bentonite, and/or or whether and or not a surface casing should be set.
Available drilling equipment	Contact local drillers and identify the general availability of hollow-stem auger, hydraulic and air rotary, and cable tool drilling rigs in the area for establishing alternatives. Inquire about borehole diameter, depth capabilities, and costs.

Initial Site Inspection

The purpose of an initial site inspection is to determine the extent of the groundwater quality or pollution. An initial inspection should determine the priority for conducting a more in-depth study and establishing a monitoring program. The information gathered during an initial inspection is gathered from inspection of the surrounding area and strata, examination of records, and discussion with on-site personnel. Information included, but not limited to, should be determined in the initial inspection in Table 6-5.

Nature of the Solid Waste. The types of soils can vary from site to site and section to section within an area and, historically, within a given location.

Area and Thickness. The size and depth of a land-strata are important parameters in determining leachate production as to volume and concentration.

Landfilling. Cover thickness, material, procedures, cell thickness, and so on, all affect leach volume and quality.

Visual Survey of Topography and Geology. Runoff, infiltration, recharge, discharge, and the general direction of groundwater flow can be estimated from the initial survey.

Preliminary Data

Preliminary investigations include a review of existing data gathered during the site inspection and from all available sources. Data gathered from other than the initial site inspection are used to plan the monitoring program. Planning information that must be gathered, once the need for a monitoring system has been confirmed, includes:

- Precipitation records for the site or a nearby area.
- Geologic and topographic maps covering the site.
- Geologic logs including existing wells or boring at or near the site.
- Aerial photograph(s) in order to prepare a base map.
- Information on other potential contamination sources, if any, including but not limited to type and volume of waste and method of disposal.

Preliminary Site Investigations. Data that will aid future investigations include:

- Analyses of area water samples from surface water bodies, existing wells, seep, and springs.
- Analyses of samples from surface leachate seeps.
- Botanical examination of site vegetation for signs of stress.
- Determination of area water usage not just as it applies to consumption but including recreation, fishing, wildlife, and so on. Table 6-5 lists requirements and uses for planning monitoring wells.

Hydrogeologic Setting

The most important determinant of a monitoring system design is probably the hydrogeologic setting. Existing sites have been located on such high contamination potential sites as abandoned gravel pits because little thought was given in the past to groundwater pollution. Therefore, the hydrogeologic setting will aid in the determination of the severity of potential contamination and, in turn, the need for monitoring. The hydrogeologic setting is defined by several factors:

- Surfacial Geology—Surfacial geology is the determination of the areal extent and thickness of the various deposits under the landfill. Permeability and interconnections of the layers should be determined, as the leachate threat to the groundwater depends on this. Surfacial geology can be determined by: (1) a review of geologic data gathered during the initial site inspection and preliminary site inspections, (2) planned geophysical studies, (3) test drilling to confirm the geophysical studies, define critical areas, and provide detailed samples.
- Bedrock Geology—Bedrock may act either as a barrier to leachate movement or, as in fractured rock zones, leachate may contaminate bedrock aquifers. Determination of bedrock geology is the same as for surfacial geology.
- Groundwater—Actually takes many forms and names: capillary water, soil moisture, connate water (water trapped in rock at time of formation), and so on. Figure 6-13 shows some of these waters. As can be seen subsurface water is divided into two major subdivisions, the unsaturated zone (also known as vadose zone) and the saturated zone. Figure 6-14 provides a more detailed picture of the groundwater profile. The saturated zone is what is commonly called groundwater, and groundwater is that water used by man for public and industrial consumption. Groundwater investigations should, as a minimum, include:

  Depth to the water table.
  Groundwater mounding, if any.
  Natural rate and direction of flow(s).
  Landfill effect on rate and direction of flow(s).
  Locate recharge and discharge area.
  Types and interconnection, if any, of aquifers.
  Rate of leachate infiltration relative to total flow.
- Existing Water Quality—Baseline data are required, both of groundwater free of contamination and of contaminated water, should the monitoring be in conjunction with a remedial action. The baseline data is required in order to understand possible secondary reaction in the groundwater and in geological formations such as cation/anion reactions, redox reactions, pH changes, and so on.

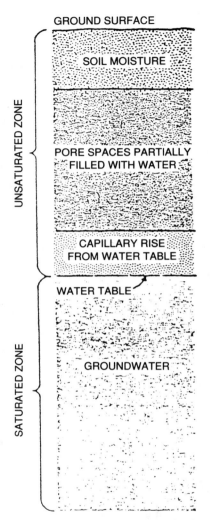

Figure 6-13 Groundwater profile. Relationship between unsaturated and saturated zones

- Determination of the Leachate Generation Rate—Is determined by a water-balance study. Data required for a water-balance study of a landfill include: precipitation data, surface characteristics, vegetation data, topography, groundwater flow, rate of landfilling, and landfill treatment practices.
- Determination of Pollution Potential—The complexity and design of a monitoring system may be greatly influenced by the pollution potential/threat. Pollution is a synthesis of the gathered data. Conclusions would be drawn from:

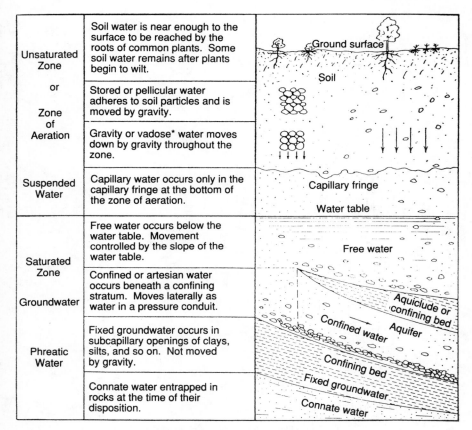

Figure 6-14 Groundwater profile detailed

- The location, size, and rate of the movement of the contaminant plume, an example can be seen in Figure 6-15.
- Affected aquifers and potentially affected aquifers.
- Types of contaminants present.
- Degree of natural attenuation effected by subsurface sediments.

With the information gathered to this point, a monitoring system can be designed. The complexity and goal of the system can be defined and monitoring locations and depths can be determined. The time frame and priority of monitoring to pollution are now decided. Once all the data have been digested, the monitoring systems can be put in place.

The most important task which must be undertaken in the development of any successful groundwater monitoring program is to first define the objectives the program is to achieve. Is the program going to be instituted solely to meet the requirements? What constituents should be considered for monitoring? What criteria

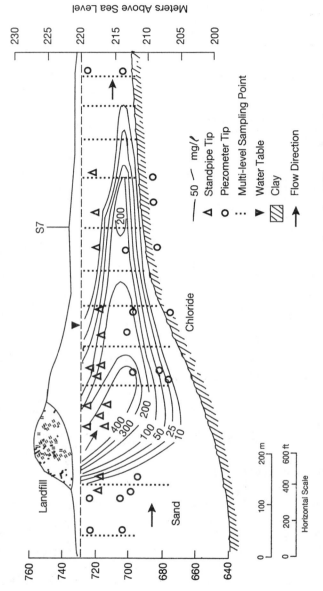

Figure 6-15 Plume of leachate migrating from a sanitary landfill on a sandy aquifer; contaminated zone is represented by contours of Cl- concentration in groundwater. (Source: Freize, R. and J.A. Cherry, Groundwater, Englewood Cliffs, NJ: 1979.)

should be used for selecting an appropriate analytical laboratory? These questions and many others must be answered before any actual monitoring may commence. Needless to say, the particular site specific characteristics of a facility groundwater hydrology will influence the amount of site investigation required to meet the program's objectives.

Often to satisfy the minimum regulatory requirements for groundwater monitoring programs, facility owners are required to install one upgradient monitoring well and three downgradient monitoring wells. The wells must then be sampled for particular parameters on a frequency mandated by requirements or the applicable regulatory program. For facilities which were in existence at the time recent groundwater protection regulatory programs were promulgated, that is, RCRA and NJPDES. Many have elected to install a groundwater monitoring system which conforms to the minimal requirements only. Unfortunately in areas with complex groundwater flow hydrology, four monitoring wells may not be completely adequate. As these facilities move from interim permitted status to full permitted status, additional wells will in all probability be required by the permitting agency.

A new facility does not enjoy the luxury of interim status. They must obtain a full permit before the regulatory agency will allow operations to commence. Depending on the particular regulatory program the facility is subject to, up to a full year of monthly groundwater sampling may be required to establish the groundwater's history, in terms of drinking water suitability, groundwater quality, and groundwater contamination indicator concentrations. The number of wells to be installed will depend on the size of the facility, the complexity of the geology underlying the area, and the amount of money available for the program over that required to satisfy the minimal regulatory requirements. The number of wells to be installed should be determined by the number which will enable the owner to establish the system's hydraulic gradients. All monitoring wells should be placed as close as possible to the potential source of contamination, for example, surface impoundments, so that leaks are discovered as early as possible.

The third type of site which requires a groundwater monitoring program is a site where groundwater contamination has been discovered. This situation is usually discovered when local water supply wells are tested, following complaints by the users. The first step in such a system is to define the extent of the contamination problem. To adequately define a contamination problem in sand or gravel, or in a relatively homogeneous aquifer, from 50 to 100 wells may be required at a cost of approximately $100,000 or higher, including chemical analyses. The wells should be screened at different depths at each well location to determine horizontal and vertical heads and differences in the water quality with depth. The key to such a program is to have enough wells and enough data so that the origin of the contamination (poor housekeeping, multiple sources, one source, or migration onto the property from off-site) can be determined.

The first step in designing a groundwater monitoring program is to investigate the history of any location or site. The information gathered from this investigation will provide clues as to the types and extent of possible contamination which will have to be considered. Potential on-site contamination sources include abandoned and active landfills, waste-water impoundments, buried product pipelines, old sewers, tanks, product storage areas, product loading areas, cesspools, dry wells, storm water collection areas, and spill incidents. The age of potential on-site contamination sources will affect the degree of contamination present and the extent of the contaminant plume migration. An inventory of the raw materials used and wastes disposed of over the life of an area should also be compiled. A review of old aerial photographs and interviews with former employees can provide information on past practices which are no longer readily apparent. Finally, any previous monitoring results should be obtained which will give some perspective on the area's historical groundwater and surface water quality.

Once a site investigation is completed, a regional inventory should be compiled to determine if the on-site contamination may be due to off-site sources. This regional inventory can be developed through field inspections, examination of land-use maps, and aerial photographs. Old aerial photographs will often be the only means available to identify landfills which have been covered and built on, lagoons which have been filled in, and industrial facilities which have been dismantled. The investigator should also review state and federal environmental and transportation agency files for spill reports and other inventories which may have been compiled by these agencies.

Background information on an area's geology and hydrology is essential for evaluating potential leachate generation and its vertical and horizontal migration through the groundwater system. Many U.S. geological survey publications as well as county soil surveys provide such background information. Other factors which may impact on an area's groundwater flow include engineering activities, such as artificially filled areas, buried trenches containing pipelines, storm drains, temporary or permanent dewatering facilities, and groundwater pumpage. These activities must be identified and taken into consideration when designing a groundwater monitoring program.

A number of nondrilling investigative methods are available which will give an idea of groundwater characteristics before the installation of the monitoring wells. Backgroundwater quality data can be obtained by sampling the area's springs, water supply wells, and standing surface water bodies. Infrared photography can be employed to pinpoint contaminated groundwater discharging to surface water bodies and areas of distressed vegetation. Measuring the area's resistivity is also a method which is useful for identifying shallow contaminated groundwater bodies. This method is only applicable where there is a significant contrast in water quality, the water table level is less than 40 feet deep, the geology of the aquifer is relatively homogeneous, and there are no local interferences such as buried pipelines, power lines, metal fences, and so on.

After all preliminary information is collected, including the system's flow direction via the use of piezometers, the site is now able to begin the actual installation of monitoring wells. The number of wells installed will be according to the objective for the system, that is, to meet the minimum regulatory requirements, to verify or eliminate the possibility of contaminants migrating onto the property from off-site sources, or possibly to investigate the overall groundwater quality beneath the site. During the actual drilling of the monitoring wells, a quality assurance program should be developed and followed. Otherwise, the credibility of the data collected may be questioned, and information that could be essential for subsequent interpretation of water quality and water-level data may be missed. For monitoring in the zone of saturation, three well-type alternatives exist which include: observation wells, piezometers, and well clusters. Well clusters are the preferred alternative because they consist of closely spaced wells drilled to different depths which provide information on the vertical distribution of groundwater contaminants. Individual single-screened wells provide only information on the areal distribution of groundwater contaminants and should be recognized as simple "preliminary" tools in many geologic settings. The location of monitoring wells should be as close as possible to the waste management area so that early detection of any leakage affecting the groundwater can be accomplished.

During the actual drilling of the monitoring wells, a one time opportunity to collect specific information about the groundwater system is presented. As the well is drilled, hydraulic head measurements at various depths and continuous water sampling will provide a profile of the aquifer's hydraulic head and groundwater quality, respectively. Electrical resistivity should also be employed to define permeabilities, porosities, and borehole characteristics. If there is little which is known about the hydrogeology of the area prior to drilling, the information collected during drilling will determine the final well depth and screen setting.

Now that the monitoring wells are installed, the actual monitoring of groundwater can begin. First and foremost, a permanent well numbering system must be established. Second, a well location map on a useable scale which shows well casing elevations should be drafted. Finally, a purging/sampling procedure must be developed and strictly adhered to. It is essential that stagnant water in the well casing be purged first, thus allowing the ambient groundwater to flow into the casing. The water which flows into the monitoring well is the groundwater which must be sampled for representative analysis results to be obtained. The actual sampling of the groundwater is perhaps the most important part of any successful groundwater monitoring program.

It is important in sampling groundwater to be certain that the groundwater collected has the same oxidation reduction and other chemical characteristics as the groundwater in the aquifer from which the water was collected. For example, if the monitoring well is open to the atmosphere, dissolved iron can be oxidized to ferric hydroxide which will precipitate in the well. Since ferric hydroxide is a scavenger of a variety of contaminants, the water collected from such a well will tend to exhibit lower concentration levels than those actually present in the aquifer.

Initial sampling of each groundwater monitoring well on a site is completed with one purpose in mind. That purpose is to determine how many well volumes must be pumped before the water being sampled is representative of the aquifer. A well volume is determined by multiplying the diameter of the well times the depth of the water in the well and then converting the number obtained to gallons. Typically, initial sampling should occur at 1, 2, 4, 6, and 10 well volumes and each of the samples should be analyzed for constituents of concern, specific conductance, temperature, and dissolved oxygen. These analyses results are plotted against well volumes pumped. The well volume at which tie analyses results begin to level off is the well volume at which future samples should be taken. The determination of the proper well volume removal number should be made at least twice each year for three years for each monitoring well being sampled.

The frequency of sampling is very specific to the type of monitoring program in place as well as the regulatory agencies involved. A site where contamination of groundwater has been documented and containment/cleanup measures have been initiated will require more frequent sampling than a site where no contamination has been found. It is also required that more frequent samples be taken when establishing the background groundwater quality at a site. Samples should be taken on a monthly basis for one year to establish the background concentration values. Subsequent sampling should take place at least quarterly so that groundwater contamination can be detected as reasonably early as possible.

The parameters to be measured on monitoring well samples should include EPA's groundwater quality parameters (chloride, iron, manganese, phenols, sodium, and sulfate), groundwater contamination indicator parameters (pH, specific conductance, total organic carbon, and total organic halogen), and the groundwater drinking water supply suitability parameters listed in Table 6-2. After background concentrations are established for the aforementioned parameters with monthly sampling for one year, the site owner may reduce the frequency of sampling and the number of parameters being analyzed. In the second and subsequent years of a groundwater monitoring program, it is common practice to sample on a quarterly basis for the groundwater contamination indicator parameters. As a precaution, it is also advisable to routinely analyze groundwater samples using a gas chromatograph (GC) scan for chlorinated hydrocarbons, such as PCBs and any herbicides or pesticides which are used at the site.

The analysis of groundwater samples should be as specified by the applicable regulatory agency and/or as stated in *Standard Methods*. In this way the acceptability of the analyses results from an accepted analytical procedural viewpoint will be assured. It is important to evaluate the reliability of the proposed sampling, handling, and analytical techniques stated in *Standard Methods* to the site specific conditions. Laboratory turnaround time between sample collection and data work-up should be less than two weeks. Once analysis results are received from the laboratory, the results should be reviewed as soon as possible to determine if groundwater contamination may be occurring.

This review consists of a statistical evaluation of the analytical results reported by the laboratory as stated in EPA's *Hazardous Waste Regulations* (40CFR Part 265 Appendix IV),

The owner or operator must use the test to determine statistically significant changes in the concentration or value of an indicator parameter in periodic groundwater samples when compared to the initial background concentration or value of that indicator parameter. The comparison must consider individually each of the wells in the monitoring system. For three of the indicator parameters (specific conductance, total organic carbon, and total organic halogen) a single-tailed test must be used at the 0.01 level of significance for significant increases over background. The difference test for pH must be a two-tailed Student's t-test at the overall 0.01 level of significance.

Since virtually all waste disposal in landfills and waste treatment in surface impoundments takes place above the saturated zone, one would expect that the immediate underlying unsaturated zone would be the zone that would be monitored. The major difficulty encountered in monitoring the unsaturated zone is in obtaining samples. Unsaturated soils most often will not yield water to a well because the moisture is held in the soil under surface tension at less than atmospheric pressure. If water from the unsaturated zone could be monitored, the site operator would be able to detect a groundwater contamination problem much more rapidly than the more common sampling of the saturated zone via monitoring wells. In general, the earlier a groundwater contamination event is detected, the lower the corrective action costs will be.

The reliability of methods for monitoring water quality in the unsaturated zone has been demonstrated; however, their actual practical application has been limited. The most common method used for sampling soil moisture is through the use of a pressure-vacuum lysimeter. This device does have several limitations, most notably the relatively small sample volume (500 ml or less) which can be obtained. Such a small sample volume limits the selection of parameters for chemical analysis. More importantly as mentioned previously, no detailed experiments on sample validity have yet been completed.

Structural Monitoring

An alternate means of protecting against leachate escape from a landfill site is through structural monitoring by determining if there have been any shifts in the underlying strata. A popular means of structural monitoring is through acoustic monitoring. Used as a monitoring technique in hazardous waste drains, acoustic emission monitoring is used primarily to determine the stability of the underlying and surrounding geologic strata. Relatively inexpensive compared to other methods of conventional structural monitoring, acoustic monitoring has been used in small hazardous waste dam applications. This monitoring is based on comparison and judgment regarding shifts in the level of acoustic emission activity. Acoustic measurement results are interpreted on a relative basis. Both qualitatively and quantitatively, acoustic emission readings from a field site can be compared with those taken earlier to assess relative stability. Increases in acoustic activity above normal are a sign of a slope stability problem, while decreases in activity signal a move toward slope stability. Finally, a large acceleration in acoustic activity can mean the possibility of slope failure.

Data collected during an acoustic monitoring activity can be quantified in units of count/minute. The geological activity, stability of the landfill site, and indicated action to be taken associated with various quantities of acoustic emissions can be shown. Using acoustic emission to monitor seepage from a landfill seems promising; however, to produce sufficient emissions, turbulent flow may be required, and may be expected only after piping has developed, but well before the piping leads to failure. Acoustic emission can also be used to monitor repairs in a dam. Relatively low activity compared to normal is an indication that site repair has been successful. Acoustic emission monitoring has been employed at various hazardous waste dam sites throughout the United States.

Biological Monitoring

Biological monitoring employs living organisms to indicate the presence of harmful chemicals in the environment. Through biological monitoring, early warning signs are detected and immediate action can be taken to repair the source of such chemicals. Biological monitoring also helps to identify the source of such pollutants and their pathway to human beings. By stopping the pollutants in their path, further damage to the environment and to human life can be avoided.

Both plants and animals serve well as biological monitors since there is potential for pollutant accumulation in both plant and animal tissues. In addition to showing the changes in the level of pollution and environmental contamination, studies performed on these tissues are useful in predicting direct or indirect effects on humans. Bioaccumulator organisms are organisms which take up a particular element from the environment and concentrate it in their tissues. Therefore, the organism's intake of the element is greater than its output. These elements are concentrated in certain areas of the organism's body such as the hair, feathers, leaves, and so on.

Biomagnification refers to the increase in the concentration of a toxic element in an organism from one trophic level to the next. For instance, such an increase may take place if a toxic element is present in the soil and is taken up with other minerals by a plant. The plant is then consumed by a rabbit who absorbs and retains the toxic element. If the rabbit is then eaten by a human being, it is likely that the concentration of the trace element in the soil where it originated is significantly smaller than in the tissue of the human who was the final consumer, assuming that human absorbs and retains the element.

On the other hand, biominification refers to a decrease in the concentration of a contaminant from one trophic level to the next. Such a decrease occurs when the contaminant concentration is low in the edible or digested parts of the prey, when the contaminant has been stored in inedible tissues such as bones, when the absorption of the contaminant in the predator's digestive system is low, when the contaminant is transformed from a highly absorbed to a poorly absorbed chemical form, or finally, when the retention time of the contaminant is short.

Regardless of whether a particular pollutant biomagnifies or biominifies, testing of biota surrounding a landfill is a good means of determining possible leaks in a leachate collection system. If contaminants known to be present in the leachate are suddenly showing up in plant or animal life nearby, it is a good indication that leachate is escaping. Rapid action will help alleviate further adverse impacts on the surrounding environment as well as potential health risks to humans and the possibility of lengthy legal litigation.

7 INDUSTRIAL AND COMMERCIAL REQUIREMENTS

Industrial water requirements must be:

- sufficient enough to take care of present and future requirements
- available at flow rates and pressure to meet all peak demands and provide adequate fire protection
- suitable quality for its various plant uses

These three requirements appear to be fairly obvious, and in selecting a site are necessary without qualification. In selecting a site for a new plant it is necessary to prove which sites are able to furnish sufficient water of a quality which can simply and economically be treated to suit its needs. Furnishing an adequate amount of water for an existing plant which has outgrown its supply, or for one faced with a dwindling supply, is not simple. For a well supply, lowering of the well pumps is usually the first step, but it is not possible to carry this out indefinitely. Spacing wells farther apart or acquiring a new well field may be more deisrable. In some cases, replenishment of the groundwater from surface sources, or even supplementing or replacing the well supply with a surface supply may be possible. If a surface supply is simply seasonally scanty, it may be required to develop a supplementary supply.

Other steps that should be taken by many industrial plants are to

(1) stop leaks and wasting of water
(2) study plant operations and find ways of reusing water
(3) treat, recover, and reuse as much of the wastewater as possible

The savings in water that can be accomplished by such means are often surprisingly large.

FLOW RATES, PRESSURES, AND STORAGE

Flow rates to be made available should always be based on peak demands and not on averages. Taps or pumps, mains, distribution systems, valves, and fixtures must be large enough to handle peak flow rates. Treating equipment on lines going directly to service should be large enough to handle peak flow rates. With equipment such as filters or ion exchangers, duplication units are usually employed so that when one unit of the system is off the line for backwashing or regeneration, the other unit or units of the battery can carry the full load. Treatment equipment should not be pushed beyond its rated flow rate.

Elevated storage tanks, floating on the line acting as equalization, are of value in smoothing out peak demands, maintaining an even pressure, and furnishing water under pressure for fire protection, especially in cases of a power failure. Ground-level storage basins or clear wells are also frequently employed to iron out peak demands and reduce the size of filters or other treating equipment which, if the basin or clear well is large enough, may be operated on the constant-rate, start-and-stop principle. Where pumping is employed, it should be backed up by standby equipment, so that it can furnish water in case of a power failure. This latter type of equipment would typically be automatically turned on by power a failure.

Peak flow rates or various water uses can vary over a considerable range. These are seldom less than twice the average flow rate (average flow rate in gpm is total gpd, divided by number of minutes in the working day) and may exceed four times the average flow rate. Estimation of the peak loads to be met is therefore necessary, and these data can be obtained from similar plants and experience. In making such estimates, it is obviously better to overdesign.

QUALITY

The quality of the water required depends on its end uses. As the tolerances for various impurities vary according to such uses, the quality of the water required in each case may differ greatly. For instance, sea water with intermittent chlorination may be quite satisfactory for some cooling purposes and yet not be suitable for boilers or for other cooling uses or wet processes.

If raw water is suitable in some cases, there is no need for treating the portion required for such uses. As for those portions required for other uses, in some cases, only one form of treatment such as softening may be needed while in other cases, specific treatment may be required.

If raw water is unsuitable for all plant uses as, for instance, if it is very turbid, has a high color, or contains iron or manganese, one central plant may be employed for eliminating these impurities. This may be followed by whatever other treatment may be required to furnish water of proper quality for its required uses. For certain uses, the water quality required may be so high that it demands practically complete removal of all of the impurities.

WATER CONDITIONING

Waters used in industrial plants may be briefly divided into the following groups:

- boiler feed water
- cooling water
- process water
- general-purpose water

When treatment is necessary to fit any of these uses, the type of treating equipment to be used depends on a number of factors, significant among which are the composition of the raw supply water and the quality required for use. The quality required for one boiler feed water, for instance, is not necessarily the quality required for another boiler feed water. Similarly, the quality required for one cooling water or one process water is not necessarily the same as the quality required for another cooling water or another process water.

Boiler Feed Water

In the case of a boiler feed water, if it is for boilers operating in the lower pressure ranges, removal of the hardness alone as may be effected by a sodium cation-exchange (zeolite) process may be sufficient. For boilers operating at higher pressures, a reduction of the total solids and alkalinity as well as removal of the hardness may be required. This may be effected by one of the hot processes, by the two-stage cold lime and sodium cation-exchange process, or by the hydrogen cation-exchange process followed by neutralization with caustic soda or, more commonly, with the effluent from the sodium cation-exchange process. For still higher pressures, not only a removal of the hardness and a reduction of the total solids and alkalinity but also a marked reduction of the silica content may be required. This may be effected by either a single-stage process using the sludge blanket-type of hot-process water softener or better by a two-stage process using the same type of water softener for the first stage and either the sodium-cation exchange process or phosphate treatment for the second stage. For boilers operating at very high pressures, a practically complete removal of all impurities may be required which can be effected by the ion-exchange demineralization process or by distillation. Also it might be added that the degree of deaeration required for all boilers, except those operating at low pressures, is so high that it demands a practically complete removal all dissolved gases.

Cooling Water

Treatments of cooling waters will differ according to the composition of the raw water and type of system:

- once through and to waste
- once through and then used for other purposes
- recirculated in an open system (with cooling towers)
- recirculated in a closed system

For some waters, possibly no treatment or only chlorination may be required, while with other waters possibly a reduction of the bicarbonate hardness by acid treatment may be necessary. Cooling water may often be treated so as to render it of suitable quality for both cooling and certain subsequent uses as, for instance, by the reduction of the bicarbonate hardness by the cold lime process or complete softening by the sodium cation-exchange process or by the two-stage cold lime and sodium cation exchange process. Possibly treatment by the cold lime process plus a small acid dosage may be required and for recirculation possibly treatment by the sodium cation-exchange process or in some cases demineralization by an ion-exchange process. For more details regarding cooling waters, refer to Chapter 8.

Process Water

The quality of water required for different processes varies over a wide range. The quality of water required for a given process presently may be quite different from the quality which had been used in the past.

Some waters may require no treatment or possibly only chlorination for use in certain processes, where others may require only a reduction of the bicarbonate hardness such as may be effected by the cold lime process. On the other hand, many processes require a water which is practically free from hardness and the treatment employed would be the sodium cation-exchange process or, in some cases, the two-stage cold lime and sodium cation-exchange process.

Regarding complete removal of impurities, there have always been some processes which demanded such a high quality of water and require the ion exchange demineralization process or distillation. Due to the high costs of distilled water, its use is limited to only those processes where it was an absolute necessity and, even then, it is used as sparingly as possible. An ion-exchange demineralization process has operating costs usually a small fraction of the costs of distillation. Uses of demineralized water have not only been applied to the processes which formerly used distilled water but also to a very large number of other processes where formerly the costs of such a high quality of process water would have been prohibitive.

General-Purpose Water

General-purpose water is for uses other than boiler feed, cooling, and wet processing. Water which is supplied for the use of personnel must be of approved quality and should also be free from objectionable tastes and odors. As far as the water for the lavoratories, showers, and so on, it should not be unduly hard. In many cases where softening is required, only the hot-water supply is softened and this may be accomplished by the sodium cation-exchange process. Here water is softened before it passes to the water heaters but, with some very hard water supplies, the cold-water supply is also either partially or completely softened. Where laundering operations are performed in the plant, all of the water used, both hot and cold, should be softened by the sodium cation-exchange process and such completely softened water is also best for many other cleansing operations. As for the water used for general flushing purposes, this seldom requires treatment.

Turbidity, Color, Chlorination, Taste, and Odor

While most groundwaters are clear and practically colorless, most surface waters at least on occasion contain turbidity and/or color. Removal of turbidity and/or color is accomplished by coagulation, settling, and filtration. For many industrial cooling waters and certain process waters, a clear and practically colorless effluent is not necessary but a reduction of the turbidity and/or color to certain tolerances may be required instead. If the coagulation and settling are carried out in a suspended solids or sludge blanket-type of equipment, such as the precipitator, filtration of the great bulk of the water may often be omitted and only that portion required for boiler feed or other special uses is filtered. Also, if a reduction of either the bicarbonate hardness or both bicarbonate and noncarbonate hardness is required, then the cold lime process or the cold lime soda process may often be carried out together with the coagulation and settling in the same equipment.

Because organic color in water differs, some high-color waters require coagulation at pH values lower than those in the cold lime or lime soda process. In these cases, if a high degree of color removal is required, the coagulation and settling are best effected by not combining them with the cold lime or lime soda process. Filtration may be required with such waters in order to secure the best results. If the quantities of water to be handled are relatively small and only a slight amount of turbidity is present, the settling tank may sometimes be omitted and the coagulant applied ahead of the filters.

Chlorination, when required, may be effected by feeding the chlorine or a hypochlorite to the water as it enters the settling tank prechlorination, by feeding it either to the effluent from the settling tank or the effluent from the filters (postchlorination) or both before and after prechlorination and postchlorination. Where rather large quantities of water are being treated, liquified chlorine gas is employed while, if the amounts being handled are relatively small, a hypochlorite may be used. Objectionable tastes and odors may be removed by the use of activated carbon, preceded in some cases by aeration. Tastes and odors can be removed by passing the water through an activated carbon filter which contains a bed of granulated activated carbon.

Iron, Manganese, Hydrogen Sulfide

Iron and/or manganese, when present as their soluble divalent bicarbonates, may be removed by aeration, settling, and usually filtration. If the pH value is above 7.0, iron oxidizes very rapidly to the insoluble hydroxide but manganese requires a higher pH value. When a hardness reduction by the cold lime or lime soda process is also required, aeration before the water enters the equipment will effect an excellent removal of both iron and manganese, especially in the suspended-solids type of equipment.

If complete softening by either the sodium cation-exchange or hydrogen cation exchange process is required, the soluble iron or manganese may be removed simultaneously with the hardness by either of these treatments. Another process which may be used if the iron and/or manganese is not more than about 1 ppm is the manganese zeolite filtration process. Iron and/or manganese in acid waters may be removed by aeration, neutralization to the proper pH value, settling, and filtration. Suspended ferric hydroxide or organically chelated iron and/or manganese may usually be removed by coagulation, settling, and usually filtration.

Most sulfur waters contain relatively small amounts of hydrogen sulfide and, if the pH value is not too high, they are usually aerated. The aerated water is then chlorinated to oxidize any residuals. Waters containing higher amounts of sulfides are not so readily treated and may require special treatments.

BOILER FEED WATER

Natural waters leave a residue of mineral matter when evaporated. Some of this residue may be a hard, rock-like scale which adheres firmly to the walls of vessels and heat exchangers in which the water is boiled.

Boiler Deposits

Scale formation in cooking pots is a nuisance but when generating steam in a boiler, scale becomes not only a nuisance but a menace. Scale has a very low thermal conductivity. The average is about 5 percent of the conductivity of steel, and some porous scales may possibly range down to less than 1 percent, forming fastest at points of greatest heat input. Scaled boilers therefore require a much greater temperature differential between metal and water than clean boilers operated at the same rating and may consequently become derated.

In mild boiler steels, if the scale is thick enough to cause a metal temperature much above 900°F, failure of the metal may be expected. Special steels can withstand higher temperatures but the safe operating temperature for any steel must not be exceeded. The increase in the temperature differential caused by 0.1-inch thickness, with a thermal conductivity of $K = 0.75$, is only 111°F at a heat input of 10,000 Btu per sq ft per hour but becomes ten times as large—, 1,110°F—when the heat input is 100,000 Btu per square ft per hour. Therefore, the old-fashioned low-pressure steam boiler with its low rate of heat input can stand a much greater thickness of scale than the modern high-pressure boiler with its much higher rates of heat input. Scales of one-fourth to one-half inch were not too uncommon in some older low-pressure boilers, while failures have occurred in modern high-pressure boilers with only a scale of 0.05-inch. Even 0.01-inch thickness of porous scale might cause failure of the boiler metal in the highly irradiated sections. It is therefore apparent that the terms *thick* and *thin*, as applied to boiler scale, are meaningless, for a scale that is thin enough for one boiler to withstand may be entirely too thick for another boiler. In actual practice, however, it has been found that the heat losses in boiler plants caused by scale are relatively small.

The terms *scale* or *boiler scale* are applied to adherent boiler deposits, while the terms *sludge* or *mud* are applied to the nonadherent deposits. Obviously, scale is much more troublesome than sludge, for sludge can be blown off or washed out while scales have to be turbined or chipped out. In certain locations, scale removal may be extremely difficult.

The principal scale and sludge formers are calcium carbonate, magnesium hydroxide, calcium sulfate, and silica. The solubility of calcium carbonate at 212°F is about 13 ppm, magnesium hydroxide 8 ppm, and calcium sulfate about 1,250 ppm (expressed as $CaCO_3$ equivalents). The solubilities of calcium carbonate and magnesium hydroxide decrease somewhat with rising temperatures, probably to about 5 ppm for the former and about 2 ppm for the latter at 392°F (210 psig), both expressed as $CaCO_3$. The solubility of calcium sulfate decreases much more rapidly with increasing temperatures. Its solubility at 338°F (100 psi) falls to 103 ppm and at 428°F (322 psi) to 40 ppm. Calcium sulfate also forms a very hard and dense scale which is difficult to remove, while scales composed chiefly of calcium carbonate are softer in nature and easier to remove. The solubilities given are for the pure substances in pure water and give merely the principles of scale formation. The presence of other salts, especially those containing a common ion, naturally affect these solubilities. Also the solubilities of calcium sulfate, in the lower temperature ranges, varies with gypsum, hemihydrate or anhydrite. Calcium carbonate, magnesium hydroxide, calcium sulfate, and silica are the primary scale-forming substances in steam boilers. Scale is more complex than a simple mixture of these substances with some alumina and iron as might be deduced from a chemical analysis of the scale. As later mentioned under silica scales, silica can occur as analcite or as calcium silicate.

Other compounds of calcium and magnesium which are followed in natural waters are very soluble and are not scale formers; however, magnesium chloride and sulfate are noted for their corrosiveness. Magnesium chloride is especially corrosive and the action appears to be rather catalytic in nature, with the magnesium chloride first decomposing into magnesium hydroxide and hydrochloric acid. The hydrochloric acid then attacks the boiler metal, forming ferrous chloride. Next the ferrous chloride reacts with the magnesium hydroxide to form ferrous hydroxide and magnesium chloride. Dissolved oxygen from an underaerated or only partially deaerated water also contributes in such attacks. If boiler salines carry some sodium alkalinity, magnesium chloride or sulfate would react with it to precipitate the magnesium as the hydroxide so that this type of attack could not occur.

Another scale former is calcium hydroxide, which obviously does not occur in a natural water and could only be introduced into the boiler through improper treatment. Calcium hydroxide has a solubility curve which decreases as the temperature rises, being about 2,400 ppm at 32°F, 890 ppm at 212°F, and 135 ppm at 392°F (210 psig)—all expressed as $CaCO_3$.

Silica, except in a few cases, was usually not a troublesome scale former in older, low-pressure boilers. In analyses of boiler feed waters, it was seldom determined separately but was usually lumped together with the minor constituents, ferric oxide and alumina, the total commonly being expressed as $R_2O_3 + SiO_2$. Silica can usually be kept from forming scale in the boilers by having the boiler feed water free from hardness, having an excess of sodium phosphate in the boiler salines, and having sufficient alkalinity present to hold the silica in solution. With hardness, however, silica can form calcium silicate scale in the boiler and, as mentioned previously, two other types of silica scales which may be formed are the double silicate scales, with alumina or iron and a base and a scale composed principally of silica. These scales are usually collectively termed *silica scales* as silica is the primary cause of their formation. Silica is also soluble or volatile in high pressure steam and so may be carried over and deposited as hard glassy scale on turbine blades.

Turbidity and sediment, if allowed to enter the boiler, may form both sludge and scale, the latter by cementation or baking on the metal or by silica contents reacting with the alkaline boiler salines. Therefore, turbid waters are usually clarified by pretreatment before softening or demineralization. This pretreatment may, in certain waters, consist of coagulation, settling, and filtration. Or when the waters contain appreciable amounts of bicarbonate hardness, the pretreatment may consist of cold lime softening, coagulation, and filtration, In the same equipment, the removal of turbidity, iron, or manganese may be accomplished simultaneously with the reduction of the bicarbonate hardness.

Boiler Feed Water Make-Up Treatment

Boiler feed water make-up is applied to the treated raw water which is required to make up the water losses from the boiler caused by use of open steam in process work or steam otherwise lost plus the water lost in the boiler blowoff. Where there are no condensate returns or none fit for reuse in the boiler, the feed water is 100 percent make-up. In large power plants, surface condensers recover so much of the evaporated water that the make-up may constitute a very small percent of the water fed to the boilers. Between these extremes are all the other cases. The percentage of make-up required in the same plant may be highly variable. Variations may be seasonal, but in some cases, the make-up requirements vary greatly from hour to hour.

Evaporation of a boiler is best stated in pounds per hour. This may be translated into gallons per hour (gph) by multiplying by 0.12. Another method of expressing evaporation is based on a unit called a boiler horsepower hour which is assumed to be equivalent to the evaporation of 4 gallons of water per hour at 100 percent of rating. If the rating is 200 percent, the evaporation is 8 gph, and so on. For example, a 250-hp boiler operated at 300 percent of rating would evaporate 3,000 gallons of water per hour.

Make-up is approximated from the total evaporation figure less the condensate returns. This is an approximation, for corrections should be made for the water lost in blowoff. In the case of all of the various water-softening processes, except the hot lime soda process, the open steam used in heating the effluents may be assumed to cancel out approximately the amount of water lost in the blowoff. In the case of the hot lime soda process, the capacity is based on the hot effluent, at that temperature about 8 lb per gallon, which is made up of the treated water, plus the steam used in heating it, so that a correction for blowoff should be added.

Sodium Cation-Exchange Zeolite Process

The sodium cation exchange process is widely used for softening boiler feed waters, especially for boilers operating at lower pressures. The chief advantages are complete removal of hardness and simplicity of operation, which may be either manually operated or completely automatic. The disadvantages are that it does not reduce either the alkalinity or the total solids contents of a water. Since sodium cation-exchanger water softeners are widely used for softening process waters for a host of various industries, it is often common practice to install one central water-softening unit of this type and use the effluent for process work, boiler feed water, and all other required plant uses.

The cation exchanger used may be one of the siliceous types such as: processed and stabilized glauconite greensand or a synthetic sodium alumino-silicate zeolite; a carbonaceous type, or a synthetic resin type. The carbonaceous type and the synthetic resin type have the advantages of being nonsiliceous in nature and lessen the possibility of silica pickup by the water passing through the exchanger bed. The original silica content of the water passes through the bed unchanged for none of the cation exchangers have any silica removal properties. If silica removal is required, this may be accomplished by the two-stage cold lime and sodium cation-exchange process or the two stage hot lime and sodium cation exchange process.

Other methods of reducing the alkalinity include:

- partial neutralization of the softener effluent with sulfuric acid
- the two-stage, sodium cation-exchange and chloride anion-exchange process
- the hydrogen cation-exchange and sodium cation-exchange process

None of these remove silica and only the ion exchange reduces the total solids content.

Hydrogen Cation-Exchange Process

The chief advantages of the hydrogen cation-exchange process are that it completely removes the hardness, reduces the alkalinity to whatever level is desired, reduces the total solids by an amount equivalent to the alkalinity reduction, is simple to carry out, and operates on cold water. Its disadvantage for treating high-silica content boiler feed waters for high-pressure boilers is that it does not reduce the silica content. In carrying out this process, if the sulfate and chloride content is very low, the train of equipment may consist of:

- hydrogen cation-exchanger unit(s)
- degasifier
- caustic soda feed

If the sulfate and chloride content is higher, the train of equipment may consist of:

- hydrogen cation-exchanger unit(s)
- sodium cation-exchanger unit(s)
- degasifier

In the first arrangement, the caustic soda is adjusted so that it neutralizes the mineral acids produced by the hydrogen exchanger and furnishes whatever excess alkalinity is required. In the second case, the flows are adjusted so that the sodium bicarbonate content of the effluent neutralizes the mineral acidity of the effluent from the cation exchanger and furnishes whatever excess of alkalinity is required. In the ion-exchange demineralization processes, hydrogen cation exchangers are used to remove the cations and anion exchangers to remove the anions.

Ion-Exchange Demineralization

Ion-exchange demineralization and silica removal are used for demineralizing boiler feed waters for high-pressure boilers. Depending on the composition of the water to be treated, the degree of demineralization and silica removal required, and the operating and capital costs, various arrangements of equipment may be employed. Considering first a two-step demineralization and silica removal system, the first step would be carried out by passing the water through a hydrogen cation-exchanger bed and the second by passing the effluent from the first step through a highly basic anion-exchanger bed. In the first step, hydrogen cations would be exchanged for the calcium, magnesium, sodium, and other cations which were taken up by the cation exchanger. This results in the formation of both strongly ionized mineral acids, such as sulfuric, hydrochloric and/or nitric acids, and weakly ionized acids such as carbonic and silicic acids.

In the second step, both the strongly ionized and weakly ionized acids formed in the first step would be taken up by the strongly basic anion exchanger which would give in exchange an equivalent amount of hydroxyl ions to unite with the hydrogen ions to form water. At the end of the operating run, the cation exchanger would be backwashed, regenerated with an acid, usually sulfuric acid, rinsed, and returned to service and the anion exchanger would be backwashed, regenerated with caustic soda, rinsed, and returned to service.

The equipment used might therefore consist of a hydrogen cation exchanger unit, or a battery of units with their respective regenerant tanks. Such systems are often used with waters which have low alkalinity or in cases where the volumes of water handled are relatively small. With waters which have appreciable bicarbonate alkalinity contents, it is more economical to use a degasifier or vacuum deaerator between the two steps to remove the carbon dioxide formed by the breakdown of carbonic acid into carbon dioxide and water than it is to remove it in the second step. Any small residuals of carbon dioxide remaining in the effluent from the degasifier or vacuum deaerator obviously are removed in the second step together with the silicic acid and strongly ionized acids.

Cold Lime Soda

Chief advantages of the cold lime soda process are that it reduces hardness to a relatively low amount from 85 ppm to about 17 ppm, depending on the excesses of chemicals used. Total solids and alkalinity are reduced by an amount nearly equivalent to the amount of carbonate hardness removed. This process also raises the pH value, reduces silica when magnesia or dolomitic lime is employed but not as efficiently as the hot process, and operates on cold water.

The cold lime soda process is used in softening municipal water supplies. This process, omitting the soda ash and using only lime and a coagulant, is used for:

- reducing the bicarbonate hardness
- reducing the bicarbonate hardness and removing turbidity
- reducing the bicarbonate hardness and removing iron and manganese

It is therefore used for conditioning cooling waters, certain process waters, and in certain cases as pretreatment in connection with ion-exchange demineralization processes. The cold lime process is also used as the first step of the two-stage cold lime and sodium cation-exchanger process as outlined in the following paragraphs.

Two-Stage Cold Lime and Sodium Cation Exchange

The chief advantages of the two-stage cold lime and sodium cation-exchange process are that:

- It softens the water completely.
- It reduces the alkalinity by an amount corresponding to the reduction of bicarbonate hardness.
- It reduces the total solids content.
- Because operating costs are low, lime treatment is usually the most economical for removing bicarbonate hardness and the sodium cation-exchange process for removing noncarbonate hardness.
- It may be used to remove silica, and it operates on cold water.

A disadvantage is the capital cost which is higher than for either a cold lime soda plant or a sodium cation-exchanger plant. Magnesia can be used in the removal of silica from boiler feed waters, to a limited extent, and for softening and removing silica from water for stationary boiler plants. It is also used to a limited extent for softening process waters and municipal water supplies. As for the removal of silica by the use of magnesia, this is much more efficiently carried out in the two-stage hot lime and sodium cation-exchange process.

Distillation

The advantages of distillation are that it removes almost completely not only the hardness but also all other mineral substances present in the raw water. Its chief disadvantages are the high initial and operating costs, so that distillation is largely confined to high-pressure power plants where the condensate returns constitute more than 95 percent of the boiler feed. In such plants, the stills or evaporators, as they are usually called, operate by degrading high-pressure steam, which usually reduces operating costs. Distilled water, to a very limited extent, is also used in certain industrial processes, but in such cases the operating costs are generally high.

Silica

A silica removal method is the ferric hydroxide process in which ferric sulfate is added to the water and the ferric hydroxide is produced by the reaction with alkalinity reducing the silica content. Rather large doses of ferric sulfate (8 to 20 ppm for each ppm of silica removed) increase the sulfate content of the water, which in the cold or hot lime soda processes means also an increase in the total solids content of the treated water. This process of removing silica is seldom used.

Another silica removal process is the magnesia process used in the cold lime soda water-softening processes. If the amount of magnesium hydroxide precipitated from the magnesium hardness of the water being treated was not sufficient to accomplish the required degree of silica reduction, additional magnesia was supplied by the use of dolomitic lime and/or activated magnesia. However, in the cold precipitation processes, it is necessary to use magnesia-dissolving equipment in which some of the sludge is recirculated into the water before it enters the sludge-blanket (suspended-solids contact) type of cold process treating equipment. A number of large two-stage cold lime and sodium cation-exchanger plants use this process of silica removal in the first stage of this two-stage process. Since magnesia removes silica much more efficiently in the hot-water softening processes which requires no dissolving equipment, this process is now seldom used.

The magnesia process of silica removal in the hot lime soda process, the two-stage hot lime soda and phosphate process, and the two-stage hot lime and sodium cation-exchange process are widely used. The hot process softener, which is used for this purpose, is the sludge-blanket (suspended-solids contact) type, as this type of softener is effective in reducing the silica content and also furnishes a much clearer effluent for the filters than the older conventional types. If the amount of magnesium hydroxide precipitated from the magnesium hardness of the water being treated is insufficient to effect the desired degree of silica removal, additional magnesia is supplied by the use of dolomitic lime and/or activated magnesia.

Silica removal in the ion-exchange demineralization processes is accomplished by the use of highly basic anion exchangers which result in an excellent removal of the silica. Various arrangements of demineralization equipment may be employed depending on the composition of the water being treated, the required results, and the operating operating and capital costs.

Deaeration

Deaeration of boiler feed waters is for the removal of oxygen, nitrogen, and carbon dioxide. Oxygen is corrosive, nitrogen is inert, and carbon dioxide is corrosive and lowers the pH value of the condensate. Deaeration, when accomplished in properly designed equipment, will lower the dissolved oxygen content to less than 0.005 ml/1 (0.007 ppm).

This is usually followed by treatment with a deoxidizing agent, such as sodium sulfite or hydrazine, to remove any oxygen residuals or oxygen which may be accidentally introduced with the condensate returns. An excess of sodium sulfite can be maintained in the boiler salines. This may range from 30 ppm at pressures up to 750 psi but may be decreased with rising pressures to around 5 ppm at pressures over 1,000 psi. At higher pressures, sulfur compounds may be given off in the steam. At very high pressures, hydrazine presents the advantages of not introducing mineral solids.

Deaeration of the condensate also eliminates dissolved oxygen. Ammonia, amines, or filming amines may also be used as corrosion inhibitors for return lines. In low-pressure boilers, dissolved oxygen contents range from a few hundredths up to 0.3 ml/l which are often tolerated. A better practice is to deaerate the water as completely as possible even for relatively small boiler plants for which packaged aerators are available.

Condensate Returns

Condensate returns are distilled water and usually contain only small amounts of mineral impurities, ranging from a few ppm to possibly slightly over 20 ppm. The hardness content may range from zero to possibly 15 ppm, depending largely on the relative tightness of the condensers and the quality of the cooling water used. Some carbon dioxide may also be present as well as some dissolved oxygen, both of which may be removed by deaeration. This may then be followed by the use of a deoxidizing agent. As for any hardness contents, in most cases this can be precipitated by internal treatment with phosphate in the boiler.

In boiler plants where the boilers are operated at very high pressures, the condensate returns should be of the highest quality. This includes not only the condensate returns to boilers operating in the supercritical range (over 300 psi) but also to many boilers operating in the so-called subcritical range and to the boilers in nuclear power plants. The condensate returns to boilers operating at these very high pressures should be treated to reduce both insoluble and soluble impurities, such as iron, copper, dust, silica, and electrolytes to minimum amounts in order to prevent deposits in boilers, superheaters, and on turbine blades and minimize corrosive effects.

Treatment of the condensate therefore involves both filtration and demineralization. Obviously dissolved oxygen residuals should be removed and hydrazine has been used to eliminate any traces which may be present in the make-up or condensate. For filtration and demineralization of the condensate, filtration is accomplished with special types of filters especially made for this purpose, as the filter medium and demineralization is effected usually in mixed-bed ion-exchange demineralization equipment. Since the amounts of impurities in the condensate passing through the demineralization equipment are very small, usually expressed in ppb, parts per billion high flow rates may be employed and the volumes of condensate treated between regeneration are large.

Internal Treatment

Internal treatment of boiler salines may be required to:

- Overcome slight hardness residuals, from externally treated make-up or that introduced by condensate returns, by adding a soluble phosphate.
- Overcome the corrosive tendencies of any residuals or accidental introductions of dissolved oxygen by adding sodium sulfite or hydrazine.
- Treat raw make-up water with boiler compounds.

In boilers operating at very high pressures, hydrazine may be used to remove trace residuals of dissolved oxygen as it does not introduce any mineral solids into the boilers. This, of course, is not the case with sodium sulfite. Its reaction with oxygen results in the formation of water and nitrogen as shown in the following reaction:

$$N_2H_4 + O_2 \rightarrow 2H_2O + N_2$$
$$\text{Hydrazine} \quad \text{Oxygen} \quad \text{Water} \quad \text{Nitrogen}$$

In such cases, ammonia may be used in the boilers to raise the pH value and this also does not introduce mineral solids.

Sulfite Treatment

As noted, dissolved oxygen residuals in make-up and condensate may be removed by feeding sodium sulfite in sufficient amounts to react with oxygen residuals and to furnish and maintain a slight excess in the boiler salines. At boiler temperature sodium sulfite reacts rapidly with oxygen to form sodium sulfate, as shown in the following reaction:

$$Na_2SO_3 + \tfrac{1}{2}O_2 \rightarrow Na_2SO_4$$
$$\text{Sodium Sulfite} \quad \text{Oxygen} \quad \text{Sodium Sulfate}$$

The concentration of dissolved oxygen is usually expressed as cubic centimeters per liter (cc/l) or milliliters per liter (ml/l), which is numerically the same quantity. It may also be expressed as ppm. If sodium sulfite is used as a deoxidizing agent, an excess is usually maintained in the boiler salines and this excess varies in amount depending on the operating pressure of the boilers.

In cases where demineralized or distilled water is employed and both feed water and condensate have extremely low dissolved oxygen content, the amounts of sodium sulfite fed may be very small and the excess may be based on the purity of the sodium sulfite and an efficiency of about 75 percent.

Phosphate Treatment

Condensate may introduce small amounts of hardness into the boiler. Although the amounts present in the condensate may be very small, the concentration caused by evaporation in the boiler may increase them to appreciable amounts. Similarly, residual hardness in externally treated make-up concentrates in the boiler salines in any make-up water.

A soluble phosphate added to the salines in amounts sufficient to precipitate the hardness and maintain an excess in the boiler salines will react with the hardness residuals and precipitate them in a nonadherent form and thus prevent scale formation. As with sodium sulfite excesses, phosphate excess used varies according to the pressures at which the boilers are operated.

Phosphates are also used in boiler compounds to precipitate the hardness of raw feed waters, usually in low-pressure boilers. In the case of very high-pressure boilers where both the feed water and the condensate are demineralized and where the introduction of mineral solids is to be avoided, ammonia may be employed to raise the pH value.

Blowoff

In all the methods of treating boiler feed waters, except demineralization and distillation, the sulfates and chlorides are not removed and appear as the sodium salts in the feed water. When these waters are used in the boiler, these sodium salts concentrate in the boiler salines. If this is continued unchecked, the concentration will reach a point at which foaming and/or priming will result. While these two terms are commonly used synonymously, *priming* refers to sudden, somewhat explosive boiling, known to chemists as bumping, while *foaming* refers to the formation of a foam consisting of very tiny bubbles of steam dispersed through the upper layers of the boiler salines. Both foaming and priming result in a carry-over of salts and water with its attendant disadvantages.

Specific concentrations at which foaming or priming will occur vary according to several factors such as:

- composition of the salines
- presence or absence of suspended matter
- design of the boiler
- pressure and rating at which the boiler is operated.

The higher the pressure and rating, the lower is the concentration at which foaming or priming will take place. As to the relative foaming tendencies of various sodium salts, this can be highly variable.

Corrosion Prevention

In the boiler, corrosion is best prevented by removing the primary causative agents, oxygen and carbon dioxide, from both the make-up and condensate returns by deaeration and by the maintenance of a high pH value in the boiler salines.

When the boiler feed water is softened by the hot lime soda process, the primary heater used will remove more than 95 percent of the dissolved air. This is usually sufficient for lower-pressure boilers operating without steel tube economizers or stage heaters. Where the latter are used and for all high-pressure boilers, total deaeration is required and, with the hot lime soda process, the deaerator is usually built into the softener as an integral part of the equipment. With other types of water softeners, a separate deaerator is employed and this is usually of the two-pass type where all of the steam used in the primary heater is first blown through its effluent, thus lowering its dissolved oxygen content to 0.005 ml/l or less.

With the hydrolysis of sodium carbonate in the boiler salines, there is the consequent formation of sodium hydroxide and the liberation of free carbon dioxide in the steam. Low alkalinities in the make-up water are desirable for reducing corrosion in condensate return lines. Corrosion by oxygen and carbon dioxide occurs only when the liquid phase is present. Dry steam lines are not attacked but the corrosion in condensate return lines can be severe, especially if any liquid is pocketed. Complete deaeration of the make-up and condensate used in the boilers and the maintenance of a low alkalinity in the boiler salines greatly reduce corrosion, but returns from systems used for heating buildings may introduce air with resultant dissolved-oxygen corrosion. Correct pitching of the return lines in such cases so that water cannot be pocketed and so that they drain rapidly will greatly reduce this type of corrosion. In some cases, amines may be used to inhibit corrosion and, as noted, in very high-pressure boilers ammonia may be employed.

Cooling Water

Cooling water used for surface condensers should not be allowed to form scale or other heat-insulating deposits, for an inefficient heat transfer with a consequent lowering of the vacuum on the discharge from the prime mover is uneconomical. Refer to Chapter 8.

COMMERCIAL AND INSTITUTIONAL WATER CONDITIONING

Office Building

Most large office buildings obtain their water supplies from the municipality in which they are located. Carefully controlled municipal water supplies are of approved bacteriological quality and are almost always clear and low in color and usually are also free from objectionable tastes and odors.

Office buildings often have their own steam boilers while others may buy steam. Where steam boilers are used, the boiler feed water should be softened. Softening equipment typically consists of pressure-type sodium cation-exchanger (zeolite) water softeners, which are usually installed in batteries of two or more units to assure continuous service. Cooling waters containing appreciable amounts of bicarbonate hardness may be softened in the same equipment and then treated with caustic sodium silicate to inhibit corrosion. If the water contains sufficient bicarbonate hardness to form scale in the hot-water system, then all of the water going to the heater should also be softened. If laundry work is done in the building, then all of the water, both hot and cold, which is to be used in the laundry should be softened. The softened water will also be found to be best for other cleaning operations. Also, both the hot and cold water for restaurants, beauty shops, or barber shops in a building should be softened and this can be accomplished by the use of relatively small, individual sodium cation-exchanger units. Corrosion in the water heater and hot-water lines may be checked by feeding caustic silicate of soda but this treatment should not be applied to the boiler feed water. Corrosion in condensate return lines may be greatly retarded by deaeration of the boiler feed water and by pitching the return lines so that they will drain rapidly without pocketing any liquid water.

Hotels

Water requirements for hotels will range from about 300 to over 650 gallons per guest room per day, an average being about 400 gallons. Hot water requirements will range from about 20 percent of the highest total to 40 percent of the lowest total. Peak flow rates will be approximately four times the average gpm for a 24-hour day. Most hotels, motels, and clubs obtain their water from a municipality. Hotels, motels, and clubs which have a private water supply use a groundwater, which usually is obtained from deep wells, or in some cases, from springs. However, some may use a surface supply. In general, most groundwaters, especially those obtained from deep wells, are usually clear, colorless or low in color, and free from harmful bacteria. The latter should not be taken for granted and bacteriological tests should be made to insure that this is always the case. Quite a number of deep well waters contain iron and/or manganese which are objectionable constituents since they stain drinking glasses and porcelain fixtures, impart an astringent metallic taste, turn tea black, (the iron combines with the tannic acid of the tea to form a dilute ink), make coffee of an unsatisfactory quality, and make laundering an impossible chore. Also some well water contains hydrogen sulfide (sulfur wates) which has an extremely disagreeable rotten-egg odor.

The amount of water used in institutional laundries is usually 3.0 to 3.5 gallons per pound of work laundered. When hard waters are used in institutional laundries the results are:

- waste soap and other detergents
- form insoluble deposits which tend to stick to the laundered materials resulting in poor quality of work with a harsh feel to it
- require excessive amounts of bleach
- shorten the life of the laundered materials

CORROSION

Corrosion, in general, is the phenomenon of the interaction of a metal with its environment, resulting in its deterioration or destruction. The environment may be water, soil, or air. Metals tend to revert through corrosion to their more stable compounds and the phenomena depends on the condition of the water, soil, and material of construction under consideration. Parameters that affect corrosion include alkalinity, hardness, pH, specific conductance, total dissolved solids, chloride, and trace metals. Metallic corrosion problems may be encountered under either oxidizing or reducing conditions and are aggravated by high dissolved-solids content.

Other materials such as plastics may deteriorate under the influence of dissolved chemical substances. Potential problems may also exist due to microbial attachment. Construction materials for pipes and tanks are available in a large variety of metallic and nonmetallic materials.

Important causes of corrosion from water are:

- galvanic corrosion
- hydrogenation
- electrolysis
- chemical reaction
- direct oxidation
- stress corrosion
- stray current electrolysis
- bacterial corrosion

Corrosion tends to reduce the strength and life of metals and therefore of structures and pipes, thus resulting in enormous monetary losses every year. Methods to protect pipes and tanks from corrosion and to control corrosion should be used.

Corrosion Protection

Protection from corrosion can be obtained through protective coating and linings. Materials used for such coating and linings are the asphaltic materials, enamels, resins, lacquers, zinc, galvanizing, plastics, paints, coal-tar enamel, vitreous coatings, cement linings, and others. Protective linings and coatings are preventive measures and must be adopted at the time of manufacture of pipes and the construction of steel tanks and structures, or during construction phases of equipment used for storage or transport of water.

Cathodic Protection

Cathodic protection is the application of electricity through an external power supply or by the use of galvanic methods for combating electrochemical corrosion. Cathodic protection should be used as a supplement and not as an alternative to other methods. Direct current (DC) power requirements vary from 0.4 to 10 kW in most cases. The main power loss occurs in the anode earthing and earthing can be carried out by any metal (pure or scrap), carbon, coke, or graphite.

Protection by Galvanic Anodes

Galvanic anodes serve the same function as the cathodic protection system but do not require continuous electric supply as in the case of cathodic protection. The required current is supplied by an artificial galvanic couple in which the part to be protected, usually steel, is made the cathode by choosing the other metal, having the higher galvanic potential, as the anode. Zinc, aluminum, and magnesium of sufficient purity or their alloys which are higher up in the galvanic series must be used for this purpose. Such anodes are generally spaced at 4-6 m along the pipe line.

Deposition of Protective Coatings

A thin protective layer of $CaCO_3$ is deposited by the water inside the surface of pipes. This is accomplished by adjusting the pH value and alkalinity of water, to keep the Langelier Saturation Index to a slightly positive value. Lime or soda ash or both can be used to raise the pH and alkalinity.

Small amounts of sodium silicate (Na_2SiO_3) will deposit dense, adherent but slightly permeable film. A dosage of 12-16 mg/l is maintained in the beginning and is gradually reduced to 3-4 mg/l.

Treatment of Water

Treatment of water like an adjustment of pH, removal of CO_2 and excess O_2, increase in calcium-ion and carbonate-ion concentrations, and addition of inhibitors can overcome to a large extent the corrosive tendency of water. Chemical treatment can be effective as only a supplement to the other methods such as protective coatings.

As stated, corrosion is the deterioration or decay occurring when a material reacts with its surroundings or the fluid being transported or contained. It may be either uniform, where the material corrodes at the same rate over the entire surface, or localized with only small portions affected. There are twelve types or classes of corrosion.

Uniform corrosion occurs over the entire metal surface at the same rate. It can be controlled by proper selection of materials and using protection methods such as coatings.

Galvanic corrosion occurs when two different metals are in contact in a conductive solution. An electrical potential exists between the different metals which serves as a driving force to pass current through the corrodent. The result is more corrosion of one of the metals in the couple. The more active metal becomes the anode and corrodes at a faster rate than the cathode. When, for example, joints and valves of two different piping materials come in contact, the possibility of galvanic corrosion exists.

Erosion corrosion occurs from the movement of a corrosive over a surface, increasing the rate of attack due to mechanical wear and corrosion. Erosion is attributed to the stripping or removal of protective surface film or adherent corrosion products. Erosion appears as smooth-bottomed shallow pits. Such an attack may also have a directional pattern which is attributed to the corrodent path moving over the pipe surface. The rate of corrosion is increased under high-velocity conditions, especially during turbulence or impingement. Fast moving slurries containing hard or abrasive particles are likely to generate such corrosion. First, of course, is the selection of a more resistant material. Erosion can also be reduced in transport applications by increasing pipe diameters which decreases velocity and turbulence. Also flared tubing can reduce problems at the inlet of tube bundles. Generally, erosion occurs sporadically.

Cavitation corrosion is a special form of erosion corrosion. Cavitation is produced by the rapid formation and collapse of vapor bubbles at a metal surface. High pressures that are produced can deform the underlying metal and remove protective films. Smooth pipes reduce sites of bubble formation and lessen cavitation corrosion.

Fretting corrosion is another form of erosion corrosion. Fretting occurs when metals slide over each other, producing mechanical damage to their surfaces. Vibration generally causes sliding. Corrosion products cause continued exposure of fresh surface that actively corrodes. The use of harder materials reduces friction.

Crevice corrosion is common at gaskets and lap joints, and comes from dirt deposits and corrosive products. This type of corrosion can be attributed to one of three things: (1) changes in acidity in the crevice; (2) lack of oxygen in the crevice; (3) the build-up of a harmful iron species or the depletion of an inhibitor. Alloys are generally less susceptible to crevice corrosion than pure metals.

Pitting corrosion is the formation of holes on an unattached surface with the share of the pit responsible for continued growth. This is generally a slow process, taking months or sometimes even years before first traces are apparent.

Exfoliation corrosion begins on a clean surface but spreads below it. The attack has a laminated appearance, and entire areas can be eaten away. Exfoliation is marked by a blistered or flaky surface with aluminum alloys most commonly attacked. It is combated by heat treatment and alloying.

Selective leaching or *parting corrosion* is the removal of one element in an alloy. An example is the leaching of zinc in copper-zinc alloys (dezincification).

Leaching is detrimental since it adds a porous metal to the effluent and is combated by utilizing nonsusceptible alloys.

Intergranular corrosion involves an attack upon grain boundaries. When molten metal is cast, it solidifies at randomly distributed nuclei and grows in a regular atomic array to form grains. This is known as the grain boundary. The atomic mismatch offers an ideal place for segregation and precipitation. Corrosion takes place because the corrosion attacks the grain-boundary phase.

Under severe conditions, entire grains are dislodged due to the complete deterioration of their boundaries. A surface that has undergone intergranular corrosion appears rough and feels sugary. The grain-boundary phenomenon that produces intergranular corrosion is heat sensitive. Such an attack is generally a by-product of a heat treatment, welding, or a stress relieving operation. The cure is another heat treatment or selection of a modified alloy.

Stress-corrosion cracking is due to the combined action of tensile strength and a corrodent. It is the most serious of all corrosion problems because many alloys will undergo stress-corrosion cracking. Stresses that cause cracking are due to residual cold work, welding, and thermal treatment, or may be externally applied by mechanical injury.

The cracks are generally in intergranular or transgranular paths, and such corrosion usually takes a long time. Preventative measures include: stress relieving, removing the critical environmental species, or proper selection of a more resistant material.

Corrosion fatigue is a special form of stress-corrosion cracking. It is caused by repeated cyclic stressing, and occurs in the absence of corrodents. It is common in structures which are subject to continued vibration. The presence of a corrodent increases susceptibility to fatigue.

Most construction materials are expected to undergo some type of corrosion. Therefore, it becomes important to determine what effects chemicals in an environmental system will have on materials. Careful analysis must be made of effluents, and existing piping and construction materials should also be examined and compared. The following factors influencing the extent of corrosion should be considered:

- concentration of major constituents being handled
- pH of effluent
- temperature of effluent
- degree of aeration (limited aeration may enhance certain types of corrosion)
- velocity of the fluid stream in the transport system
- inhibitors
- startup and downtime procedures

Construction materials for pipe and tanks are available in numerous materials, metallic and nonmetallic. Physical properties should be thoughtfully examined before any final selection is made.

Corrosion and Corrosion Control

Useful water is water that is noncorrosive that will cause neither deterioration of metals in distribution or domestic plumbing or industrial systems, nor "red" water or reduction of flow capacity in the piping. Many natural and treated waters are corrosive; therefore, corrosion prevention is often a necessary part of water supply practice or appropriate materials of construction should be employed. Water is corrosive to a metal if it dissolves the metal or furnishes substances that react with it at the metal-water interface.

Corrosion or dissolving of metals is a natural phenomenon related to the fact that metals normally occur in their oxidized forms in nature; for example, iron ores are usually found as oxides. These oxidized forms are chemically reduced by mineral refining processes to metal which is unstable in the presence of oxygen and water and tends to return by corrosion to the more stable oxides.

Typically the corrosion of iron is characterized by pitting which results from the solution of iron and the deposition of ferric hydroxide and other reaction products (tuberculation). Red water is a common result of corrosion. The gradual solution of the metal may weaken it to the point of destruction; tuberculation reduces the hydraulic capacity of a pipe. Soft waters high in carbon dioxide are generally corrosive toward copper and may cause blue-green stains on enamel bathroom fixtures. Corrosion of aluminum produces an impervious, white aluminum oxide (Al_2O_3) deposit on the metal. Soft waters high in carbon dioxide may corrode lead pipe forming toxic lead salts.

The discussion of corrosion in this book is limited to the interior of metal pipes, tanks, and so on. Exterior metal surfaces may be corroded by exposure to certain soils. The resulting corrosion may be either electrochemical or biochemical due to the action of anaerobic bacteria. Nonmetals such as concrete and cement-asbestos plastics in contact with water are subject to a deterioration that is sometimes called corrosion.

Corrosion of metals exposed to water results from the formation of minute electric or corrosion cells on the surface of the metal. Each cell consists of an area in which metal atoms are dissolving (the anode) and an area in which the excess electrons that have flowed from the anode react with constituents of the water (the cathode). These areas with their different electric potentials are caused by variations in the chemical or physical properties of the metal or water from point to point or time to time.

The solution and entry of positive metal ions into the water at the anode is an oxidation process that must be accompanied by a reductive process at the cathode. Oxidation releases electrons that must be captured by reduction in order to maintain the electrical neutrality of the water.

These reactions can continue only if the excess electrons flow because of the net difference in electrical potential and corrosion potential from the anode to the cathode where they enter into one or more reactions that may occur there simultaneously. Certain of these reactions may result in the deposition of reaction products, for example, ferric hydroxide (rust). The anodic and cathodic reactions that occur in any particular corrosion cell depend on the nature and concentrations of the dissolved ions. The reactions prevailing at any time in a given system will be those which in combination produce the greatest electromotive force, namely the anodic reaction with the highest potential and the cathodic reaction with the least potential.

Corrosion therefore is an electrochemical process, and it depends on chemical reactions and on the flow of electric current. Essentially corrosion takes place in three steps:

- Atoms of a metal in contact with water go into solution as ions, leaving an excess of electrons on the metal in anode area.
- These electrons flow through the metal to the cathode area.
- In the cathode area, they react either with hydrogen ions in the water to produce hydrogen gas that collects on the metal surface, or with other ions dissolved in the water to form corrosion products that may be deposited as insoluble precipitates in the cathode area.

In summary, corrosion may be viewed as a combination of two types of processes:

- Anodic processes (oxidation) in which metal atoms dissolve to form ions; for example, $Fe \rightarrow Fe^{++} + 2\ e^-$. That is, the metal yields ferrous ions plus two electrons.
- Cathodic processes (reduction) in which the excess electrons react with other ions to remove hydrogen ions from the water, $2H^+ + 2\ e^- \rightarrow H^2$, or with other ions in the water to form corrosion products; for example, $Fe^{++} + 3/2 H_2O + 3/4 O_2 + e^- \rightarrow Fe(OH)_3$. That is, ferrous ions react with oxygen, water, and electrons to produce ferric hydroxide (rust) as shown in Figure 7-1.

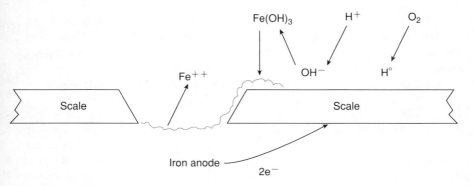

Figure 7-1 Typical electro-chemical corrosion cell

Factors Influencing Rate of Corrosion

Corrosion of a metal can occur only in situations where the metal is in contact with water containing dissolved substances. The corrosiveness of a particular water depends on so many interrelated factors that no single index or equation can correctly predict either the rate of corrosion of a specific metal exposed to that particular water or the success of a proposed corrosion control method.

Although the estimation of corrosion rates in the laboratory is extremely difficult, a knowledge of the factors that significantly influence corrosion rates is basic to the choice of corrosion control procedures.

The corrosion rate of a metal corresponds to the rate at which the anode reactions take place and is measured either in terms of weight loss as grams per year or penetration as inches per year. Theory indicates that the rate of corrosion depends on the rate of flow of electric current from the anode area to the cathode area of the metal. Since reactions at the cathode are normally slower than those at the anode, they control the corrosion rate.

The rate of corrosion of a specific metal exposed to a particular water depends on certain properties of the metal and certain characteristics of the water, such as:

- The relative position of the metal in the electromotive or galvanic series which is a measure of its tendency to be oxidized, that is, to lose electrons and go into solution. Zinc, magnesium, and iron have relatively high solution pressures and so tend to corrode; copper and silver have relatively low pressures.
- The relative ability of the metal to form insoluble compounds that may act as protective coating, for example, aluminum, and zinc.
- Joined dissimilar metals.
- Stress differentials, differences in microstructure, impurities or inclusions, in a single metal.
- Nonuniform oxide scales or deposits on the metal.

The term *galvanic corrosion* is sometimes used to designate corrosion involving two dissimilar metals. All aqueous corrosion is galvanic in nature. As we have seen, corrosion does not require the presence of two different metals. Corrosion caused by joined dissimilar metals is also termed *bimetallic*.

Water characteristics that significantly influence the rate of corrosion of a metal include the amount of dissolved oxygen, since the rate of corrosion of a metal is usually controlled by the reactions at the cathode. These are essential to the formation of hydrogen gas at the cathode. This reaction and the overall corrosion reaction cannot proceed unless the hydrogen gas film is continually removed by depolarization. At pH values above 5 reaction with dissolved oxygen is the only means of removing the hydrogen: $2H^+ + 2e^- + 1/2O_2 \rightarrow H_2O$. For a given set of factors, the dissolved oxygen concentration or more specifically the rate at which dissolved oxygen is supplied to the cathode governs the rate of corrosion.

The hydrogen-ion concentration (pH) of a water governs the single electrode potential of a metal exposed to it, thus influencing the types of reactions that occur at the cathode and the anode. The pH value also controls the solubility and hence the precipitation of various substances such as ferric hydroxide and calcium carbonate. A pH of 10 favors the deposition of the latter.

The concentration of carbon dioxide in a water affects the bicarbonate equilibrium and thus the solubility of the carbonates and film formation. Water can contain considerable amounts of carbon dioxide without being corrosive if the CO_2 is in equilibrium with the bicarbonates and the carbonates.

The absolute and relative concentrations of other inorganic ions in the water, particularly calcium, bicarbonate, chloride, and sulfate, may favor or retard corrosion by either promoting or hindering the formation of protective films. Calcium, bicarbonate, phosphate, and silicate ions tend to form insoluble anodic products, thus reducing corrosion. Chlorides and sulfates enter into reactions that yield soluble products which may interfere with or even break down protective films.

Increasing temperatures tend to increase the speed of chemical reactions in general and the rates of the reactions responsible for corrosion approximately double for every 25-30 degree Fahrenheit rise in temperature up to about 160 degrees. Temperature also influences the solubility of corrosion products. Calcium carbonate is one of the few exceptions to the general rule that increasing the temperature of the water increases the solubility of the compound. Temperature rise also decreases the solubility of gases in water.

The rate of flow of the water past the metal surface governs the rate at which the dissolved oxygen essential to corrosion is replenished at the metal surface. Conversely, high velocities may scour away undesirable corrosion products as well as furnishing a constant supply of ions that inhibit corrosion, such as phosphates. Turbulence associated with high velocities increases the rate of diffusion of oxygen to the metal surface.

Corrosion control methods commonly used in water supply practice involve choices that include:

- A choice of nonmetallic materials or corrosion-resistant metals in construction.
- Nonmetallic materials include asbestos-cement, reinforced concrete, and plastic. Examples of corrosion-resistant metals are aluminum, stainless steel, nickel, silicon, copper, brass, and bronze.
- The choice of metallic coatings, such as zinc (galvanizing) or aluminum, to protect metals.
- The choice of nonmetallic coatings to protect metals. Such materials include coal tar enamels (bituminous), asphaltics, cement mortar, epoxy resins, vinyl resins and paints, coal tar-epoxy enamels, inorganic zinc silicate paints, and organic zinc paints.

- Choice of chemicals for the water treatment. Included are deposition of protective coating or film on the metals by use of calcium carbonate, sodium hexametaphosphates, and silicates; removal of oxygen; removal of free carbon dioxide; pH adjustment.
- Electrical control (cathodic protection).

Very rarely is one of these methods used by itself as a satisfactory means of controlling corrosion. For example, chemical control methods should always be regarded as supplements rather than as substitutes for the proper choice of metals and mechanical coatings. Cathodic protection is usually tried only after other techniques have failed. Hence, in most water supply systems we find some combination of one or more of the preceding methods.

The following sections, except that dealing with cathodic protection, describe chemical corrosion control methods which the plant operator can employ as a part of water treatment. It is assumed that the choice of materials and of mechanical coatings will have been made already by the engineer who designed the water supply system.

The addition of chemicals to water in order to deposit a protective coating or film is the most important practical means of controlling corrosion. In the pH range normally encountered in waterworks practice, the corrosion rate is controlled by the reactions of oxygen (primarily oxygen depolarization) at the cathode. A substance which promotes film formation when added to water is known as an inhibitor. Cathodic inhibitors reduce the rate of corrosion by preventing dissolved oxygen from reaching the cathodic areas, thereby inhibiting the rate-controlling cathode reactions. Anodic inhibitors are less effective and not commonly used. Inhibitors that are known to make certain metals more passive or resistant to corrosion are called passivators; nitrites and chromates passivate iron.

Following are several methods widely used in water treatment practice to control corrosion by film formation that inhibits cathodic reaction.

Calcium Carbonate. The deposition of a thin calcium carbonate film on a metal protects it from corrosion. Although neither the chemical reactions responsible for the deposition nor the manner in which the film acts are well understood, calcium carbonate is thought to act as a cathodic inhibitor. A water in which all three forms of alkalinity and carbon dioxide are in equilibrium is said to be saturated with calcium carbonate: $CO_2 + HOH \ H_2C_3 \ H^+ + HCO_3 \ H^+ + CO_3^= + Ca^{++} \ CaCO_3$ (solid). These relationships are shown in Figure 7-2. Such water will neither dissolve nor deposit calcium carbonate, that is, it is stable with respect to this substance, and also toward iron.

There are several different measures of saturation or stability of which the Langelier Saturation Index is the best known. Several other indices based on the same equilibrium relationship have been proposed, namely, the driving force index, the momentary excess, the Ryzner index; they are discussed in appropriate textbooks. Unfortunately saturation of water with calcium carbonate does not

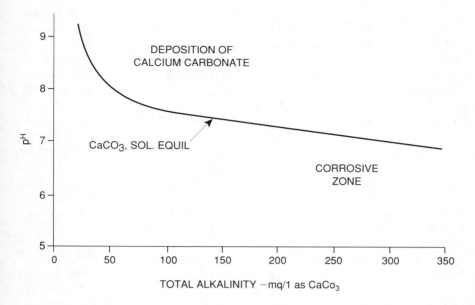

Figure 7-2 Relationship between pH and alkalinity or calcium carbonate saturation

always assure noncorrosiveness; consequently, this index measures a trend or a tendency. Other factors besides the deposition of calcium carbonate, such as pH, alkalinity, and dissolved oxygen, play important roles.

Controlling corrosion by producing a stable water warrants consideration in the plant to measure its effectiveness for specific water. Lime, caustic soda, or soda ash can be added until the alkalinity and the pH of the water plotted on Figure 7-2 indicate that the water is saturated with calcium carbonate.

The Enslow stability indicator is a device in which the water is passed slowly and continuously through a column of granular calcite or marble. Passage through the column is the equivalent of the marble test with the effluent representing the saturated or equilibrium sample.

Sodium hexametaphosphates (polyphosphate) and complex chemicals $(NaPO_3)_n$ have been used as corrosion inhibitors. The compounds react with iron oxides in the water to produce substances which by electrodeposition form a film that is bonded to the cathode, increasing its degree of polarization. Polyphosphates also show a sequestering action that complexes or ties up a variety of ions including iron, manganese, and calcium, thus preventing their deposition as rust or scale. Red water and tuberculation can sometimes be prevented in this way. The success of this method, sometimes known as threshold treatment, in a particular situation cannot be predicted, but should be determined by trial and error. Commercial products are available as plate-like, lump, flake, or granular materials but are relatively expensive. They are usually dissolved on a batch basis and fed as solutions at the rate of not more than 5 milligrams per liter. A variety of other inhibitors, organic and inorganic, are commercially available but are not approved for use in drinking water.

Removal of Oxygen. Although the rate of corrosion is in most cases proportional to the concentration of dissolved oxygen, its removal is not a practical means of control in waterworks practice. The exposure of water to the atmosphere during and after treatment tends to keep water saturated with oxygen. Oxygen, like carbon dioxide and other gases, can be removed from water by deaeration in which the water in the form of a spray is exposed to a partial vacuum. Molecular oxygen leaves the water temporarily but returns as soon as the water is exposed to the atmosphere. Deaeration is used in boiler water practice where smaller volumes are used and more careful control is possible. Oxygen may also be removed by deactivation through passage of the water through beds of some form of iron with which it will react. This method has little or no applicability for public water supplies.

Removal of Carbon Dioxide. The influence of carbon dioxide on corrosion depends on its concentration relative to the carbonates and bicarbonates in the water. It tends to hold the carbonates in solution as bicarbonates, thus interfering with the deposition of calcium carbonate. Aggressive carbon dioxide, that in excess of the amount in equilibrium with carbonates and bicarbonates, may be reduced by aeration to about 5 mg/l as a lower limit. Carbon dioxide can be neutralized by the addition of lime, caustic soda, or soda ash to any desired level. The presence of carbon dioxide in the air sets the lower limit for removal by aeration and makes it difficult to maintain a lower level in any water exposed to the atmosphere.

Adjustment of pH. This is merely another approach to the calcium carbonate equilibrium method, rather than a separate method. Adjustment of the pH, usually raising it to the equilibrium value shown in Figure 7-2, has been discussed. Lime is the cheapest way of increasing pH, but it increases the hardness of the water which may be undesirable in waters already containing significant hardness. Lime is difficult to handle and feed and often causes scaling and clogging of feed lines. Soda ash adds no hardness, is more easily fed, but may be deposited in feed lines. Caustic soda, available as a liquid, is more expensive than either, but adds no hardness and is not deposited in feed lines.

Electrical (cathodic) protection as previously described is used in controlling corrosion, but electrical or cathodic protection is not be regarded as a substitute for methods previously discussed, namely, choice of materials and coatings and chemical control by means of water treatment. Cathodic protection is achieved by the application of an outside electric current to the metal in order to reverse the electrochemical corrosion process. An auxiliary anode is added to the corrosion cell or battery and acts as a substitute for the original anode with the metal being protected now acting as the cathode. The auxiliary may be either energized by an external source of direct current or may actually be a metal higher in the electromotive force series than the metal being protected. Metallic ions from the substitute anode, rather than from the metal being protected, go into solution. The auxiliary anodes may be expendable, such as aluminum, magnesium, or zinc, or nonexpendable, such as carbon or graphite. Sufficient direct current must flow to the metal to maintain metal to water potentials of 0.80-0.85 volts.

Cathodic protection is most often used in water supply systems to protect metal storage tanks.

Iron and Bacteria

Among the types of bacteria which develop in water containing iron, one of the most commonly found is Crenothrix. Crenothrix needs iron for food. These organisms live in the dark and will not grow well in the presence of large concentrations of dissolved oxygen. When water contains iron, large masses may be present in the mains and cause reduction in flow. Objectionable concentrations of iron will precipitate in the water and death of the organisms will create disagreeable tastes and odors. Removal of iron from the water and an increase in the dissolved oxygen by aeration will prevent a growth of Crenothrix. Copper sulfate or chlorine in doses higher than used in the regular treatment of water supplies may be used to kill the organisms. Expert advice should be sought before treatment is started for the destruction of the organisms in a distribution system.

The distribution system and its components should be planned, designed, and constructed to prevent the introduction of foreign substances and minimize reactions between the water and elements of the system. Control over design and construction should take into account the following:

- Water mains should be separated from sewers by a horizontal distance of at least 10 feet, and by a vertical distance of at least 18 inches.
- Distribution system reservoirs and pump stations should be made inaccessible to the public. Standpipes and tanks should be covered and protected against surface drainage. Reservoirs should be covered and lined whenever the investment is justified. Valves and drains should be designed to exclude surface drainage.
- Cross connections must be prevented. All plans and specifications for new construction should be reviewed by competent individuals to eliminate the possibility of backflow of polluted water either as a result of backpressure or backsiphonage. There shall be no physical connection between the distribution system and any pipes, pumps, hydrants, or tanks which are supplied from or contaminated from any source.
- Mains should be sized and interconnected in such a way as to promote circulation and to minimize dead ends.
- Hydrants should be located to permit flushing of all mains.
- Air-relief and vacuum release valves should be installed. They should be

- located and protected to prevent the introduction of pollution. Valve chambers should be protected against flooding or freezing.
- Blowoff valves should be located at low points to permit the removal of accumulated sediment.
- Line (gate) valves should be located in the distribution system to permit the isolation of relatively limited areas or sections of mains during repairs.
- Materials and coatings for use in mains, tanks, and reservoirs should be corrosion resistant or should minimize corrosion. Materials and coatings that may impart taste, odor, color, turbidity, or toxicity to the water must be avoided.
- Main installation practices, including the grading of trench bottoms and joint construction that will minimize operation maintenance and repair costs, should be adopted and followed by the department.
- Mains should be protected against pollution during construction and installed in such a way as to exclude the possibility of subsequent pollution.
- Newly constructed mains, tanks, pumps, and all other components of the distribution systems should be disinfected as described later before being put in use.
- Each addition to the distribution system should be recorded accurately, exactly as constructed, on a master map of the system.

An operating and maintenance program should be executed that will achieve the distribution system objective of delivering to the customer adequate quantities of water of satisfactory quality at the least total annual cost on the goal of any supply system. Such a program should include the following elements as are applicable to a particular system:

- Maintenance of a measurable chlorine residual in all parts of the system at all times. This practice minimizes the growth of bacterial and fungal slimes in the system.
- Disinfection of all newly constructed, repaired, or recently shut down mains, tanks, and other components of the distribution system, according to the following procedure.
- Regularly scheduled and systematic flushing and/or cleaning of all dead-end mains and other mains subject to the deposition of solids. In some cases, routine flushing of all mains may be desirable.
- Regularly scheduled and systematic inspection, cleaning, and painting of all reservoirs and tanks.
- Regularly scheduled and systematic inspection and testing of all hydrants and valves in the system, normally twice a year.
- Algal and weed control in open reservoirs where required.
- Avoidance of reservoir drawdown so great as to cause vortexing above the outlet.

- Avoidance of sudden changes in direction or in the velocity of flow in the mains.
- Periodic surveys of plants and premises in which cross connections are likely to exist and the elimination of such connections.
- Maintenance of a record of the type and location of all backflow prevention devices.
- Cathodic protection in cases where mains or tanks cannot be protected against corrosion by either water treatment or by coatings.
- Maintenance of a current record of all additions and repairs to the system.
- In cases where corrosion of the system is a problem, a recording of the physical appearance of mains, valves, and other fittings removed from the system.
- Maintenance of a record and map showing the date and location of all water quality complaints.

Satisfactory water quality in the distribution system can be assured through continuous monitoring of the quality of the water entering the system at various points in the system. An effective sampling program is essential. The most accurate measure of the quality of the water at any point in the distribution system is obtained by the examination of samples drawn directly from the main.

8   COOLING WATERS

Water is extensively used for various cooling purposes in industry and commerce. This may be for cooling condensers in power plants, distilleries, oil refineries, or chemical plants; for cooling internal combustion engines in diesel power plants, gas engines in compressor or pumping stations, or gasoline engines in automotive or aviation plants; for cooling furnace fronts in steel mills; for cooling compressors in refrigeration or liquid gas manufacturing plants; for cooling chemical and other products; for air conditioning; and for a variety of other cooling processes. The amounts of water required for cooling vary considerably, depending on the temperatures of the cooling water and the particular application. The source of the cooling water may be groundwaters, surface waters, or sea water.

Water has experienced extensive use in cooling operations, both because it is an excellent cooling medium and because it is available and inexpensive. All natural waters do, however, contain dissolved solids, gases, and a variety of suspended matter in different amounts. These contaminants can be the source of varying operating problems. Bicarbonates and sulfates of calcium sodium and iron are the most common dissolved solids. The amount of each will depend on their abundance in the earth at the source of the water. Carbon dioxide is the most common of the dissolved gases found in water and the highest concentrations exist in waters from shallow wells and lakes due to decay processes. Suspended solids may consist of silt and a variety of organic constituents. All water systems are capable of developing algae and slimes in varying degrees if environmental conditions are proper.

The presence of suspended and dissolved matter can lead to precipitation under the proper conditions causing severe scaling or fouling problems in process equipment and distribution systems. High concentrations of suspended matter can result in erosion. Both problems can be translated into lost dollars because of costly maintenance and downtimes for equipment replacement. In the case of organic constituents, the ideal environment for microorganisms to grow can exist, posing serious health hazards in certain applications.

Because of the inherent problems of scale formation and potential health hazards in some applications, cooling tower water treatment must be carefully considered in the overall cooling system design.

Cooling-Water Supply

Groundwaters. Water from wells is greatly favored for once-through cooling purposes on account of its even temperature. In general, the water from wells or springs is usually remarkably even in temperature. This is true for 30- to 60-feet deep wells in the neighborhood of some 2° to 3°F above the mean average annual temperature of the atmosphere of the surrounding area and increasing about 1°F for every 6 feet of additional depth.

Surface Waters. Surface waters are usually subject to large seasonal variations in temperature. In general, with rivers of the United States the difference between winter minimum and summer maximum water temperatures will range from 30°F to even as much as 60°F. Winter temperatures of 33° to 36°F are common with river waters over much of the country, and summer temperatures of 75° to 80°F are common and may exceed 85°F. Lakes, ponds, and reservoirs are also subject to seasonal variations in temperature which are particularly wide in range in the shallower waters.

Sea Water. At locations where sea water is available, the amounts obtainable are obviously dependent only on the size of intakes, pipe lines, and pumps. Therefore, it is used only once through and then discharged to waste. The annual temperature variations encountered in various locations may cover a range as low as 11°F to slightly over 50°F. Location and depth of intake, whether in a bay or fronting the open ocean, currents, and so on will affect the temperature. Minimum winter temperatures may be as low as a couple of degrees Fahrenheit below the freezing point of fresh water, and maximum summer temperatures may exceed 80°F.

Municipal Supply. Water for cooling purposes may be drawn from the municipal supply and is of drinking water quality. Such water supply is typically too expensive for once-through use and requires recirculation through cooling towers for reuse.

PROBLEMS INHERENT TO CONTAMINANTS

In addition to the common minerals absorbed from the soil, natural waters can also be affected by industrial drainage, often resulting in acidic conditions. Faulty processing equipment may introduce a variety of contaminants such as oils, fats, acids, alkalies, and hydrocarbons directly into the cooling system. Undesirable airborne contaminants, such as hydrogen sulfide and acid vapors released by processing equipment and fly ash from coal-burning furnaces, may be drawn into the tower and dissolved in the circulating water. Without proper control, the presence of any of these materials may cause corrosion of metal parts, wood deterioration, or loss in thermal performance throughout the entire cooling system.

There are five types of cooling-water problems encountered in cooling-tower systems. These are scale formation, corrosion, organic (algae and bacteria slime) growths, suspended matter (sand, mud, silt, and so on), and oil leakage. With the exception of oil leakage, these problems can be controlled to a certain extent by standard water treatment techniques. Different types of treatment have been employed with various degrees of success. Treatments employed include using a circulating system with a small quantity of treated make-up water, with or without the addition of chemical inhibitors to the circulating water. Another method of controlling contaminations and scale formation is the use of alloy tubes. Unfortunately, this is an expensive solution and often one that is difficult to justify economically.

In addition to the first four problems, which are of water origin, oil leakage into the water also causes problems. Oil will interfere with other treatments employed. It is therefore desirable to eliminate oil leakage as much as possible by repairing leaks as soon as they develop.

Due to evaporation, salts contained in the water tend to concentrate and could precipitate, causing scale in the system. The scaling tendency of the circulating water can be controlled by an appropriate blowdown to lower the salt content, and by the addition of treating chemicals. These chemicals are inhibitors that prevent precipitation from occurring. The corrosion problems encountered in evaporative cooling water systems concern the cooling circuit (that is, exchangers, piping, and so on) and the cooling tower itself.

Oxygen, carbon dioxide, and various chemicals used to reduce scaling can cause corrosion. Corrosion control is provided largely by the use of inhibitors such as chromates, polyphosphates, silicates, and alkalies.

Corrosion within the tower itself is due mainly to the particular conditions existing therein (air, humidity, and temperature) and also to the chemical treatment of the water. All construction materials exposed to these conditions must be selected carefully. Hardware and piping for distribution headers have been successfully made with hot-dipped galvanized steel, cadmiumcoated steel, stainless steel, and silicon bronze.

Scaling and corrosion are related phenomena. The properties of water influencing both are the calcium hardness, alkalinity, total dissolved solids, pH, and temperature. Theoretically, the preceding conditions can be controlled so that the water is in equilibrium and neither corrosion nor scaling results. In practice, however, this equilibrium is difficult to achieve since it is a border condition, and a delicate balance must be maintained.

Water corrosion of iron and steel is simply oxidation of the metal forming iron oxide by galvanic action. The rate of oxidation is faster at higher oxygen concentrations. This is why corrosion is more of a problem in recirculating cooling-water systems than in once-through systems. Likewise, the rate of attack is higher for waters of higher acidity because a low-pH water is a better electrolyte. Therefore, increasing the pH up to the equilibrium point decreases corrosion. However, increasing the pH further causes scale formation.

The principal scale-forming material in cooling systems is calcium carbonate, which has a solubility of about 15 ppm and is formed by the decomposition of calcium bicarbonate.

Scaling results when the solubility limit of calcium carbonate is recalled, at which point precipitation onto tube surfaces occurs. The extent of calcium carbonate precipitation is a function of the composition of the water and the temperature. The alkalinity, dissolved solids, and pH determine the scaling characteristics. Decreasing the pH by the direct addition of acid or by carbonization will decrease the scaling tendencies of the water within limits. If a water is on the scaling side of equilibrium, increasing the temperature will increase the scale deposition.

Calcium carbonate scale is objectionable because of the resistance offered to heat transfer in heat-exchanger equipment. It is of interest to point out that increasing the temperature will increase the rates of scaling and corrosion. Scaling and corrosion are not likely to occur simultaneously, although it is possible owing to temperature differences in the system. The water could be on the corrosive side at the inlet and on the scaling side at the outlet. If water is close to this equilibrium condition, the rates of corrosion and scaling would exist but are likely to be very small.

Corrosion is less of a problem in once-through cooling-water systems, where the oxygen content is relatively low. Likewise, scaling is less of a problem in once-through systems than in recirculating systems because the water has not been concentrated, as in the recirculating case. In recirculating systems, the water is reaerated in cooling towers, which makes it more aggressive from the standpoint of corrosion. The purpose in using circulating, cooling systems is to conserve make-up water; however, this is difficult to achieve because of the dissolved-solids concentrate build-up from evaporation losses.

Scale and Deposits in Cooling Systems

Cooling systems may become clogged with

- scale
- corrosion deposits
- sediment
- organic growths

Scale. The principal scale-forming material in cooling systems is carbonate formed by the decomposition of calcium bicarbonate into calcium carbonate, carbon dioxide, and water, as shown in the following reaction.

$$\text{Ca(HCO}_3)_2 \rightarrow \text{CaCO}_3 + \text{CO}_2 + \text{H}_2\text{O}$$

Calcium bicarbonate → Calcium carbonate + Carbon dioxide + Water

In water free from carbon dioxide, calcium carbonate has a very low solubility—about 15 ppm at 32°F and about 13 ppm at 212°F. In water saturated at atmospheric pressure with carbon dioxide at 32°F, 1,690 ppm of $CaCO_3$ will dissolve, forming calcium bicarbonate. If this water were boiled for a sufficient length of time, the calcium bicarbonate would decompose to calcium carbonate. The carbon dioxide would all be driven off, and since the solubility of calcium carbonate is only 13 ppm at 212°F, the amount of calcium carbonate deposited would be 1,620 — 13 = 1,607 ppm which is equivalent to 6.7 tons of scale and sludge per million gallons. Cooling waters, however, are not saturated with carbon dioxide, do not contain 1,620 ppm of calcium alkalinity, and are not heated to the boiling point. Very few cooling waters will have calcium alkalinities of more than one fourth of this maximum figure of 1,620 ppm and most cooling waters would have calcium alkalinities under 300 ppm. The amounts of scale that such waters can form in cooling systems are, however, very large and can be troublesome even though they never reach more than a fraction of this 6.7 tons per million gallons. Table 8-1 shows solubilities of calcium and magnesium compounds in water.

With any specific content of free carbon dioxide there is an equilibrium at each temperature, which establishes the maximum amount of calcium bicarbonate that can be held in solution. Raising this temperature decreases this maximum solubility, the preceding reaction proceeding toward the right until the equilibrium for the new temperature is reached. From this, it is apparent that with certain waters high in calcium bicarbonate content and low in free carbon dioxide, even a very slight elevation in temperature may be sufficient to form scale.T

The Langelier index (or calcium carbonate saturation index) is of value in predicting the scale-forming tendencies of a cooling water. In order to calculate the index it is necessary to have:

- the methyl orange alkalinity
- the calcium hardness
- the total solids (approximately)
- the pH value
- the temperature to which the water will be raised. This calculation is further discussed in this chapter.

Magnesium carbonate in water free from carbon dioxide has a solubility of about 100 ppm, expressed as $CaCO_3$, at 32°F and about 75 ppm at 212°F. Since it is roughly about five or six times as soluble as calcium carbonate, and since also the magnesium content of natural waters is usually much lower than the calcium content (an average figure would be that calcium hardness would constitute about

Table 8-1 Solubilities of the Bicarbonates, Carbonates, Chlorides, and Sulfates of Calcium and Magnesium

Name	Formula	(ppm of $CaCO_3$)		(gpg of $CaCO_3$)	
		at 32°F	at 212°F	at 32°F	at 212°F
Calcium bicarbonate	$Ca(HCO_3)_2$	1620	decomp.	94.5	decomp.
Calcium carbonate	$CaCO_3$	15	13	0.9	0.8
Calcium chloride	$CaCl_2$	336,000	554,000	19,600	323,000
Calcium sulfate	$CaSO_4$	1,290	1,250	75.3	72.9
Magnesium bicarbonate	$Mg(HCO_3)_2$	37,100	decomp.	2,170	decomp.
Magnesium carbonate	$MgCO_3$	101	75	5.9	4.4
Magnesium chloride	$MgCl_2$	362,000	443,000	21,100	25,900
Magnesium sulfate	$MgSO_4$	170,000	356,000	9,920	20,800

two thirds and magnesium hardness one third of the total hardness), it is evident that magnesium plays a very minor role in scale formation in once-through cooling systems. A small amount may come down with the calcium, and in recirculating systems, the concentration of magnesium alkalinity can result in contents exceeding the solubility of the carbonate.

Magnesium bicarbonate is also much more soluble than calcium bicarbonate. At 32°F, in water saturated at atmospheric pressure with carbon dioxide, its solubility is over 20 times as great as that of calcium bicarbonate. Even relatively small concentrations of free carbon dioxide will hold very much greater quantities of magnesium bicarbonate than calcium bicarbonate in solution.

Calcium sulfate has a solubility of about 1,290 ppm at 32°F, 1,540 ppm at 110°F, and 1,250 ppm at 212°F expressed as $CaCO_3$. In natural waters, in a once-through system, calcium sulfate is not a scale former. When cooling waters are recirculated, care should be taken to "blow off" enough water to keep the concentration of calcium sulfate well below 1,200 ppm expressed as $CaCO_3$, to prevent hard, sulfate scale from forming in the cooling system. This applies especially to recirculated, acid-treated waters.

All of the sodium salts, the chlorides of calcium and magnesium, and the sulfate of magnesium are extremely soluble, ranging from about 60,000 ppm to over 500,000 ppm. They are not scale forming unless concentrations are carried to extreme lengths.

Corrosion Deposits. The commonest form of corrosion in ferrous metal vessels is dissolved-oxygen corrosion, caused by the oxygen content of dissolved air. This form of corrosion is greatly accelerated by low pH values. In waters of low alkalinity and high free carbon dioxide content, the attack is much more rapid than is the case with waters high in alkalinity and free from or low in carbon dioxide content.

Attack is greatly accelerated by increases in temperature, so that water which may exert but a relatively slight attack in the cold-water lines may badly attack the metal of the cooling system. Usually this attack results in the formation of tubercles rising above each pit in the metal. These greatly reduce the available flow rates through the cooling equipment as well as oppose resistance to the passage of heat to the cooling water. Another form of corrosion is that exerted by sulfurous waters, with the formation of metallic sulfides. Partial removal of some of the sulfides by aeration may render the water more corrosive, so that it is best with sulfur waters to chlorinate after aeration to assure complete oxidation of the sulfides. Sulfide deposits are usually black and adherent but frequently shell off and clog valves and narrow pipes.

Sediment. Turbid waters containing either coarse sediment or fine suspensoids are objectionable in most cooling systems, but one notable exception is surface condensers for steam. In these surface condensers equipped with straight, small-diameter tubes, it is common practice to use certain muddy river waters without treatment; under these conditions, the tubes usually have brightly polished interiors because of the scouring action of the suspended matter at the relatively high velocities of flow employed. In other cooling systems, turbid waters are very troublesome and form clogging deposits.

Organic Growth. Iron and manganese bacteria may become troublesome in cooling systems which contain iron and/or manganese, as they form masses which slow flow rates. Such masses frequently break loose and may completely block passages. Algae growths may become troublesome in cooling towers and ponds. Delignification of wood in the cooling towers may be caused by growths of fungi. Sulfur bacteria often grow profusely in aerators and ponds handling sulfur waters.

Grasses, sponges, "pipe moss," bryozoa, molluscs, and slimes are other clogging organisms found in cooling systems. Some of the organisms form compact deposits that, at first sight, appear to be hard, inorganic scale. In sea water lines, in addition to slimes, grasses, bryozoa, sponges, and so on, shellfish such as mussels, barnacles, and oysters are especially troublesome.

Cooling-Water Types

There are four general classifications made on the basis of conditions of use, as follows:

- once-through-and-to-waste cooling system
- once-through-and-then-used-for-other-purposes cooling system
- open recirculating cooling system
- closed recirculating system

Once-Through-and-to-Waste System. This is widely used where a plentiful supply of water is available. This may be from deep wells or from a surface source such as a large river, lake, pond, or reservoir, or at tidewater locations it may be sea water. In any case, since the water is to be used once and then wasted, the treatment, if required, should be a cheap one.

Once-Through-and-Then-Used-Again Cooling System. In these cases, the treatment used should be one that not only renders the treated water suitable for cooling purposes but also makes it suitable for subsequent uses. These later uses will vary with different industries. In some, where the heat balances are favorable and large amounts of process steam are used, the subsequent use may be for boiler feed water make-up. Subsequent use of the water may be largely for process water. Also in some industries the cooling water is used first for a cooling system where a low-temperature water is required and then one or more times for cooling at successively higher temperatures before final use in boilers or for process water. Such systems, where feasible, not only result in a very efficient use of the water but also effect substantial savings in heat units, thus resulting in economics in fuel.

Open Recirculating Cooling System. The most prevailing recirculating system is one in which the cooling water circulates through the cooling system, is cooled through a cooling tower or spray pond, and is then recycled. In theory about 1 percent of the water is evaporated for each 10°F of cooling effected in the cooling tower or spray pond. Losses, however, are higher, due to drift. The water lost as spray may amount to less than 0.3 percent with induced draft or forced draft towers, and under 1 percent in deck towers. In spray towers, losses may range from 1 to over 3 percent and in spray ponds from 2 to over 5 percent. Another loss is that due to blowoff off of recirculated water to keep the dissolved solids from concentrating beyond specified limits.

With such recirculated cooling-water systems, after the initial start the only water requiring treatment is the make-up and this usually is less than 10 percent of the water recirculated. Treatment required should be based not only on the composition of the raw water but also on what the composition of the treated water will be after it has been concentrated by reuse.

Closed Recirculating System. Closed recirculating systems are employed for cooling diesel and other types of internal combustion engines and for various other cooling uses. The cooling water, after use, passes through either a water-cooled or air-cooled heat exchanger. Theoretically, no make-up is required, but in practice, a small amount is typically necessary. Recirculated water should be treated for the initial start and whatever small amount of make-up is needed. Where a water-cooled heat exchanger is employed, the water used for this purpose may or may not require treatment.

Summarizing the preceding discussion, the dissolved mineral matter in most natural waters consists mainly of calcium in the form of bicarbonate or temporary hardness, and chlorides and sulfates as permanent hardness. The tendency of the water to deposit scale when made alkaline by heating or to attack metals corrosively depends on the balance of these various constituents.

The scale formed under moderate temperatures is usually due to temporary (bicarbonate) hardness being converted into calcium carbonate, which occurs upon heating or upon an increase in alkalinity sufficient to result in calcium carbonate saturation. The solubility of calcium carbonate also affects corrosion since the alkalinity of dissolved carbon dioxide in the water is greatly reduced as the saturation equilibrium is approached. Ideally, at equilibrium the various forms of carbon dioxide (free CO_2, bicarbonate, and carbonate) are so balanced that they cause neither scale nor corrosion.

TREATMENT OF COOLING-WATER SYSTEMS

The prevention of scale formation and corrosion is common to all heat-transfer equipment, not just cooling towers. The need for protecting metal surfaces against corrosion in cooling-water systems is essential to achieving maximum system efficiency and equipment life. Corrosion that is inadequately controlled can lead to irreversible equipment damage and costly unscheduled unit outages for cleaning operations or equipment replacement. Unscheduled shutdowns can seriously undermine plant efficiency and productivity.

Effective corrosion control programs are essential in reducing unit downtimes. To be effective, the program must address not only specific corrosion problems but anticipate and prevent them as well. Consequently, effective pretreatment in addition to other corrosion control measures is important.

Corrosion control of metal surfaces depends on the formation and maintenance of a protective corrosion inhibitor film on the exposed metal surface. This protective film may be established during normal application of a corrosion inhibitor program; however, there will be some lag time before the film is completely built up. Metal surfaces that are exposed to the cooling water before the film is completed may become candidates for accelerated corrosion during the initial system operation. Normally, localized corrosion or pitting is common during these early stages of operation.

Allowing a unit to undergo no treatment for periods of time before being placed into operation can result in severe damage to exposed metal surfaces. In addition to the loss of the metal and shortened equipment life, voluminous and porous corrosion by-products may form and actually act as a barrier to the formation of the protective inhibitor film.

In dealing with metal water-cooling systems, today's trend is toward the use of nonchromate-based treatment chemicals. Nonchromate applications rely on less tenacious films for corrosion protection rather than conventional chromate systems. As such, it is extremely important that the corrosion protection film be established very early in the operation.

We can define pretreatment as the initial conditioning period whereby a corrosion inhibitor is applied to the metal surfaces of the cooling system. Pretreatment conditions must be conducive to the rapid formation of the protective barrier. The conditioning procedure should involve (1) the cleaning and preparation of metal surfaces and (2) the actual application of higher than normal inhibitor concentrations.

The cleaning and passivation can be done separately or in a combined step. There are several procedures that can be employed to clean metal surfaces. Common techniques include hydroblasting, treatment with a mild inhibited acid cleaner and/or alkaline cleaner, and the use of special surfactants during cleansing. The system must be flushed thoroughly after the cleaning stage to minimize undue metal attack by residual concentrations of cleaning chemicals.

Chemical passivation should be started as soon as possible after the cleaning of metal surfaces. Accumulation of new corrosion products can occur if it is not initiated soon after cleaning. It may be achieved by either on-line or off-line treatment of the equipment.

On-line passivation involves elevating the corrosion inhibitor concentration as high as three times normal maintenance levels. At higher concentrations, the rate at which the protective film forms is accelerated. This in turn reduces the degree of initial corrosion on clean but unprotected metal surfaces. The rate at which corrosion protection takes place depends on the temperature, pH, and inhibitor used.

Off-line passivation involves treatment of equipment currently out of service. Treatment levels are typically higher; consequently, passivation is completed more quickly. Passivation of nonchromate treatment generally uses either a polyphosphate, zinc, molybdate, or other nonchromate-based inhibitor in combination with various surface-active cleaning agents. The passivation solution should be disposed of after the pretreatment stage, rather than dumped back into the cooling system where the potential for fouling can exist due to the precipitation of pretreatment compounds such as zinc or phosphate. Table 8-2 outlines both on-line and off-line pretreatment procedures.

Table 8-2 Pretreatment Procedures

On-line Pretreatment Procedures

Increase inhibitor concentration to two to three times its normal level.

Circulate the high inhibitor concentration slurry for 4-12 hours. Maintain pH between 6 and 7 at temperatures between 49° and 60°C. If ambient temperature must be used, increase the pretreatment period to 24-48 hours.

After passivation, the system should be deconcentrated. Reduce the inhibitor concentration to normal maintenance level.

Initiate normal treatment program.

Off-line Passivation Procedures

Thoroughly clean system of all dirt, oil, scale, organics, and corrosion by-product.

After system cleaning, refill with fresh water. Add the pretreatment formulation to the required concentration level.

Circulate the solution throughout the unit, maintaining pH levels between 6 and 7.

Circulation should continue for 2-12 hours at temperatures between 49° and 60°C.

After passivation, remove the pretreatment solution and replace it with normally treated cooling water.

Place unit back on line and resume normal service.

The first methods of cooling-tower corrosion control involved adding several hundred parts per million of sodium chromate, as chromate is capable of excellent anodic corrosion control at these dosages. However, these early programs were both inefficient and expensive. The advent of synergized zinc chromate-polyphosphate treatments not only made corrosion control more effective, but also lowered its cost. Excellent corrosion control requires only 30-60 ppm of inhibitor instead of a concentration one to two orders of magnitude higher. Polyphosphates are also used in cooling systems to attain sufficient corrosion control. Cooling towers are operated in a pH range of 6.0 to 7.0 to provide optimum stability for the polyphosphate. The feasibility of cooling-tower operation at higher pH levels, in which the potential for corrosion is decreased, has increased the popularity of low-chromate programs.

Cooling waters may also be treated by one or more of the following processes:

- coagulation, settling, and/or filtration
- cold lime
- sodium cation exchange
- two-stage cold lime and sodium cation exchange
- demineralization
- acid process
- iron and manganese removal, and the use of chlorination, copper salts, and polyphosphates

Coagulation, Settling, and/or Filtration. The removal of turbidity may be accomplished by coagulation, settling, and filtration. As discussed, various combinations are possible. Sedimentation, for instance, may precede the coagulation or, on the other hand, if a slight turbidity is not objectionable, sludge-blanket type of settling equipment may be used and the filters eliminated.

Coagulation, settling, and filtration may be used either alone or preceding the acid or sodium cation-exchange processes. With the cold-lime or lime and sodium cation-exchange processes, the turbidity can be removed together with an alkalinity reduction in the cold lime softener. The usual train of equipment for a coagulation, settling, and filtration plant is:

- settling tank
- coagulant feeder (plus possibly an alkali feeder and/or a clay feeder)
- filters

In addition, a sterilizing agent such as a hypochlorite (which can with suspension-type lime feeders be fed with the lime) or liquid chlorine (which would be fed from a chlorinator) might be required.

Cold Lime Process. The cold lime process is widely used on recirculating cooling systems but seldom on a once-through-and-to-waste system. The chemicals used are usually lime to reduce the bicarbonate hardness, a small dosage of alum as a coagulant, frequently followed by a small dosage of sulfuric acid to adjust the alkalinity of the effluent to the desired amount. Soda ash is usually not used, as reduction of the noncarbonate (or sulfate) hardness is an unnecessary expense for a recirculating system. In a once-through-and-then-used-for-other-purposes system, reduction of the noncarbonate hardness may be required, in which case soda ash or the two-stage cold lime and sodium cation-exchange process may be used.

With the older types of cold lime softening equipment, filters are almost invariably used. When the sludge-blanket type of softening equipment is used, the filters are often omitted. In such cases, the train of equipment usually consists of (1) a sludge-blanket type of cold lime water softener, (2) a lime feeder, (3) an alum feeder, and (4) an acid feeder. If the water contains iron or manganese or is high in free carbon dioxide content, an aerator may precede the softener, often being mounted above it.

Sodium Cation-Exchange Process. The sodium cation-exchange process is often used on a once-through-and-used-for-other-purposes system and on recirculating systems but not on a once-through-and-to-waste system. It is very widely used for furnishing cooling water for diesel or gas engines in power plants and pumping and compressor stations, as well as in other industries.

In addition to its practically complete removal of hardness, the sodium cation-exchanger water softener has advantages in its compactness and extreme simplicity of operation. The sodium salts formed by exchange with the cations of calcium and magnesium are extremely soluble so that, unlike the acid process, blowoff is not based on the sulfate content and higher concentrations may be carried. This means that smaller amounts of treated water are wasted.

In the case of unaerated, clear, deep well waters which contain iron or manganese, the softener will remove these as well as the hardness. In the case of turbid surface waters, the softener should be preceded by coagulation, settling, and filtration.

Two-Stage Cold Lime and Sodium Cation-Exchange Process. The water is first treated with lime in a cold lime water softener and then the clarified effluent is passed through a sodium cation-exchanger water softener. This process has the advantage of producing a completely softened water of low alkalinity and with less total solids than if softened by the straight sodium cation-exchange process.

This process is most applicable to a once-through-and-then used-for-other-purposes cooling system where a completely softened, low alkalinity, and low total solids water is desired. It is seldom used on a recirculating system and is too expensive for a once-through-and-to-waste system. The train of equipment used in carrying out this process usually consists of:

- cold lime water softener
- lime feeder
- coagulant feeder
- acid feeder or recarbonator
- filters
- sodium cation-exchanger water softener

Demineralization. The ion-exchange demineralization process is used only for closed recirculation systems where a very high-quality cooling water is required.

It is used for the cooling water for certain electrochemical processes, for the cooling water for supercritical pressure boilers, and so on. Demineralization systems used may be one of several types; the widely used type is a system which consists of:

- hydrogen cation-exchange unit(s)
- weakly basic anion-exchange unit(s)
- a degasifier, decarbonation tank, or a vacuum deaerator

With some water, the third item may be omitted. The preceding system does not remove silica so with cooling systems where silica removal may be required, a system consisting of (1) hydrogen cation exchange unit(s), (2) degasifier or vacuum deaerator, and (3) strongly basic anion exchange unit(s) maybe used. If the bicarbonate content of the water is low or if the amounts of water to be treated are relatively small, item (2) may be omitted. Mixed-bed units may also be employed or one of the other systems.

Acid Process. The acid used in the acid treatment is almost invariably sulfuric acid. Hydrochloric acid has also been suggested, but since its average cost on an equivalent basis is about four times that of sulfuric, it has been used only to a limited extent.

The acid treatment is based on the fact that the sulfates of calcium and magnesium are very much more soluble than their carbonates. Calcium carbonate, the principal scale former in cooling systems, has a solubility of only 15 to 13 ppm at 32° to 212°F; calcium sulfate has a solubility of not less than 1,250 ppm, expressed as $CaCO_3$ over the same range. Although magnesium carbonate has a solubility of 100 to 75 ppm and magnesium hydroxide that of 17 to 8 ppm, magnesium sulfate has a solubility of 170,000 to 356,000 ppm (9,900 to 20,800 gpg) over the same range—all results being expressed as $CaCO_3$.

In a recirculating cooling system if the bicarbonates were changed to the sulfates by the addition of sulfuric acid, no scale would be formed if the blowoff practiced were sufficient to keep the concentration of calcium sulfate safely below 1,250 ppm (73 gpg). In practice, of course, not all of the alkalinity is neutralized with acid. A small amount of calcium alkalinity is maintained so that a very thin calcium carbonate scale will form to inhibit corrosion.

The preceding treatment is for a recirculating system. However, if cooling is a once-through-and-to-waste system, a very much smaller amount of acid is required as the treatment is carried out under pressure. Thus, the solvent ability of the carbon dioxide released by the action of the acid on the bicarbonates is retained instead of the carbon dioxide being lost to the air as in a recirculating system. In a once-through-waste system the sulfuric acid treatment is usually the cheapest treatment.

In recirculating systems, depending on the composition of the water, the sulfuric acid treatment may be cheaper or more expensive than the lime treatment. The latter is often the case with a high bicarbonate water, as the blowoff with such waters to prevent calcium sulfate scale may be too high.

Iron and Manganese Removal. If the water is an unaerated clear, deep well water any iron and/or manganese content can be removed simultaneously with the hardness by the sodium cation-exchange process. In the cold lime process or the two-stage cold lime and sodium cation-exchange process, iron and/or manganese may be removed in the first step by aerating the water before it enters the cold lime treatment equipment. If nothing but iron and/or manganese must be removed, then aeration plus the addition of an alkaline material to increase the pH value to the proper point, settling and filtration are all that is required for waters containing ferrous and/or manganous bicarbonates. With waters containing organic (chelated) iron and/or manganese, coagulation, settling, and filtration are required.

Aeration is used in the removal of iron and/or manganese present as their divalent bicarbonates. Aeration may also be practiced for the reduction of the sulfide content of sulfur waters and for the reduction of high free carbon dioxide contents.

Chlorination and Copper Sulfate Treatment. Chlorine is very widely used for inhibiting organic growths in cooling systems. To a lesser extent, copper sulfate is used to inhibit algae growths in cooling ponds and basins. In many cases, both treatments are used with the copper sulfate, as always, being used intermittently and the chlorine more or less steadily.

Applying chlorine in recirculating cooling systems, if used continuously and not intermittently, should be fed to the recirculating water rather than to the make-up water, for in the latter case, chlorine residuals obviously would not long be maintained during recirculation. In some cases, it is an advantage to feed the chlorine in more than one location and prechlorination as well as postchlorination are often practiced.

Sometimes chlorination alone may be all that is required but in most cases chlorination is used in addition to other treatments. In inhibiting certain organic growths, such as shellfish, in sea-water cooling systems, the most effective method of dosing has been to chlorinate heavily to a 0.5-, 0.6-, and occasionally 0.7-ppm chlorine residual for short periods two to four times a day. These periods, in some plants, have been three to four in number per four hours, and of 15 to 10 minutes duration for each period; in others, to periods of one hour each per 24 hours are employed.

Corrosion detection plays an important role in any corrosion control program. Most of the methods employ nondestructive test methods and include hydrogen evaluation, radiography, dynamic pressure, corrosion probes, strain gauges, and eddy current measurements. Of these, the methods employed in cooling-tower practice are hydrogen valuation and corrosion probes.

Hydrogen evaluation is used to detect corrosion in closed systems at low or slightly elevated temperatures in aqueous environments. Sensitive detectors are available to detect the presence of hydrogen, which is a by-product of most aqueous corrosion processes. This method cannot locate the corrosion but can predict the approximate total corrosion rate.

Corrosion probes detect and measure the amount of corrosion occurring at a given point in a system and can be used to estimate the total amount of corrosion and the type of corrosion anticipated. Probes are available for use in a wide variety of temperature and pressure conditions.

METHODS OF EVALUATING COOLING-WATER INHIBITORS

There are three methods available for evaluating cooling-water inhibitors. These are laboratory methods, service exposure, and sample exposure.

Laboratory Methods

In general, these methods are unreliable and often give misleading results.

Actual Service Exposure

The most reliable evaluation can be obtained by this method; however, it can be very expensive and usually only a few materials can be evaluated.

Exposing Sample Materials

This method involves the assembly of several specimens in the form of a corrosion test spool. Test specimens are weighed before and after exposure in the actual service where data are urgently needed.

Unfortunately, there are no commonly accepted standard procedures for securing reliable corrosion data. Any data collected from any test methods will have value only if they can be interpreted properly. Many corrosion environments can vary widely from day to day and even from hour to hour. Even slight variations in operating procedures can drastically affect the corrosion characteristics of the cooling tower. Therefore, it is important to establish methods on how the corrosion data are to be accumulated, evaluated, and put to use.

LANGELIER AND RYZNAR EQUATIONS: SATURATION AND STABILITY INDEX

A convenient method of interpreting water analysis for the purpose of determining the calcium carbonate solubility equilibrium conditions is embodied in the Langelier equation. The Langelier equation can be used to determine the carbonate stability or corrosive properties of a cooling water for a specific temperature when the contents of dissolved solids, total calcium, total alkalinity, and pH values are known.

The Saturation Index is the difference between the actual measured pH and the calculated pHs at saturation with calcium carbonate:

Saturation Index 1 = pH (actual) - pHs (Langelier equation)

where
$$pH = (9.3 + A + B) - (C + D)$$

where

A = total solids, ppm
B = temperature, °F
C = calcium hardness, expressed as ppm $CaCO_3$
D = alkalinity, expressed as ppm $CaCO_3$

If the Saturation Index is 0, water is said to be in chemical balance. If the Saturation Index is positive, scale-forming tendencies are indicated. Finally, if the Saturation Index is negative, corrosive tendencies are indicated.

The Ryznar equation was developed to provide a closer correlation between the calculated prediction and the quantitative results obtained in the field.

Stability Index = 2 pH - pHs (actual) (Ryznar equation)

For the preceding equation:

1. If the Stability Index is 6.5, the water is scale forming.
2. If the Stability Index ranges from 6.5 to 7.0, the water is in a good range.
3. If the Stability Index is 7.0, the water is corrosive.

The optimum value for the Stability Index is 6.6.

However, these convenient indices must serve strictly as guides rather than as absolute control methods, the reason being that uneven temperatures exist throughout a cooling system. Because of this, some exchangers will scale, some will be protected, while still others will corrode.

Note that the pH (actual) is the log of the hydrogen-ion concentration.

ORGANIC GROWTHS

Organic matter also aggravates scaling and fouling conditions in cooling systems by combining with silt and/or calcium carbonate to plug up or scale equipment, thus reducing the effectiveness of the heat-transfer surface. Microbiological growths on heat exchangers retard cooling, cut plant efficiency, and increase maintenance cost. Iron- and sulfur-reducing bacteria are often a direct cause of corrosion. Algae growths can occur in all types of heat exchangers. Chlorine and chlorinated organic compounds are the most commonly used chemicals to prevent attack from bacteria and algae. Fungus and other forms of microorganisms can biologically attack the wood inside the cooling tower. This problem can be minimized by the use of impregnated cooling-tower lumber.

Undissolved solids or suspended matter plug up cooling and condensing equipment as well as fill up the cooling tower with silt and mud, which can lead to pumping problems. In addition, suspended matter aggravates scaling conditions in cooling water because silt and mud combine physically with the calcium carbonate to produce a thicker and softer scale than would be formed by calcium carbonate alone. This interferes with heat transfer and water flow. Normally, these are eliminated by continuous filtration.

LEGIONNAIRES' DISEASE

Following the American Legion Convention at a Philadelphia hotel in July 1976 the public first became aware of a new type of disease (Legionnaires' Disease). Of the 221 cases of Legionnaires' Disease, 34 resulted in death. Since that date, the Center for Disease Control in Atlanta, Georgia has isolated and confirmed that a bacterium microorganism had produced the illness in Philadelphia and at least eleven other locations.

The Legionnaires' Disease organism was discovered breeding in cooling water at several locations where the disease broke out. The conditions for this bacteria to turn lethal were created when the energy crisis of 1973 imposed conservation measures on water usage.

More and more water is now being recycled. Reducing cooling-tower blowdown increases the volume of suspended solids, mineral and salts, and as the cycles of concentration become higher, the pH and nutrient levels for biological growth increases. The temperature of hot-water return to the tower, normally around 120-130°F, creates an ideal condition for the breeding and rapid reproduction of different organisms.

The Legionnaires' Disease organism may exist in the groundwater and even in the air. The bacterium is in a dormant state until ideal life conditions appear. Once the bacterium enters the environment of the cooling tower, it can reproduce very rapidly by binary fission, creating a potential disease outbreak. It can be carried away by blowdown or windage droplets, and unless the water treatment expert develops a comprehensive program to minimize the possibility of the bacterium breeding in the cooling system, another outbreak is inevitable. Biocides are required and must be monitored continually to ensure that the proper rate is maintained in the cooling water.

WATER ANALYSIS AND TREATMENT

For control of scaling, corrosion, and algae and bacterial growth, the cooling-tower water supply must be analyzed and properly treated. Water analysis covers three areas: water hardness, alkalinity, and detection of inerts. Hardness can be distinguished in the following:

- Carbonate Hardness. This is the presence of calcium (Ca) and magnesium (Mg) carbonate bicarbonate.
- Noncarbonate Hardness. This is the presence of other salts of Ca and Mg.
- Total Hardness. This is the sum of the carbonate and noncarbonate hardness.
- Temporary Hardness. This is the presence of Ca and Mg bicarbonate and can be eliminated by boiling the water to transform bicarbonate into insoluble carbonate. Temporary hardness is slightly different from carbonate hardness because it does not take into account the presence of carbon, which is only slightly soluble.
- Permanent Hardness. This is Ca and Mg residue in the water after boiling and differs from noncarbonate hardness because it also measures the carbonate remaining in solution.

The normal units used to measure water hardness are:

- French degrees = $gCaCO_3/100$ l water
- German degrees = $gCaO/100$ l water
- English degrees = $gCaCO_3/$ imp. gal water

The value of the different degrees are:

- 1° French = 10 ppm $CaCO_3$
- 1° German = 1.78° French = 17.8 ppm $CaCO_3$
- 1° English = 1.43 French = 14.3 ppm $CaCO_3$

With respect to total hardness, make-up water can be classified as follows:

	Total Hardness (ppm $CaCO_3$)
Very Soft	15
Soft	15-50
Medium Hard	50-100
Hard	100-200
Very Hard	200

Alkalinity is a measure of the concentration of all electrolytes that give basic reactions when hydrolyzed in water, that is, salts of strong bases and weak acids (hydrates, carbonates, bicarbonates, phosphates, silicates, borates, sulfites, and so on). Chlorides and sulfates do not contribute to alkalinity. The evaluation of alkalinity is made by titration and the results are reported in ppm of $CaCO_3$.

Sometimes the alkalinity is reported as cc of the acid (HCl or H_2SO_4) used for the titration of 100 cc of water. To convert to ppm of $CaCO_3$, use the following relationship:

$$\text{cc of 0.1 N acid} \times 50 = \text{ppm of } CaCO_3$$
$$\text{cc of 0.02 N acid} \times 10 = \text{ppm of } CaCO_3$$

The number of cc's of 0.1 N acid used for titration of 100 cc of water is frequently referred to as millivalents.

Other analysis data needed for the make-up water are pH, suspended solids (ppm), chlorides (ppm Cl), sulfates (ppm SO_4), and silica (ppm SiO_2).

It is interesting to note that when the total alkalinity is less than the total hardness, then calcium and magnesium are present in compounds other than carbonates, bicarbonates, and hydrates. In this case the amount of hardness equivalent to total alkalinity is the carbonate hardness; the remainder is the noncarbonate hardness.

The solubilities of the more common salts at approximately 120°F are:

	Chloride (ppm)	Carbonate (ppm)	Sulfate (ppm)
Sodium (Na)	270,000	290,000	310,000
Magnesium (Mg)	270,000	125	330,000
Calcium (Ca)	520,000	17	2,200

When the number of concentrations of the circulating water is in the order of 3-7, some of the salts dissolved can exceed their solubility limits and precipitate, causing scale formation in pipes and coolers. The purpose of the treatment of the cooling water is to avoid scale formation. This is achieved by the injection of sulfuric acid to convert Ca and Mg carbonates (carbonate hardness) into more soluble sulfates. The amount of acid used must be limited to maintain some residual alkalinity in the system. If the system pH is reduced to far below 7.0, it would result in an accelerated corrosion within the system. As stated earlier, scale formation and/or corrosion tendency is defined by the Saturation Index (Langelier Index) and Stability Index (Ryznar equation).

If the Saturation Index is positive (which implies that the Stability Index is less than 6.5), then the water has scale tendency and the addition of sulfuric acid in appropriate quantities would be required to prevent scaling formation. The following example illustrates the estimation of the required amount of acid.

Example 1

A cooling tower is operating with the following make-up water composition:

Ca Hardness, ppm	$CaCO_3$,	85
Mg Hardness, ppm	$CaCO_3$,	33
Total Hardness, ppm	$CaCO_3$,	118
Total Alkalinity, ppm	$CaCO_3$,	19
Sulfates, ppm	SO_4,	20
Chlorides, ppm	Cl,	19
Silica, ppm	SiO_2,	2

It is clear that the total hardness is greater than the total alkalinity. Assume, for instance, that the number of concentrations in the circulating water to reduce blowdown is maintained at 5. Also, assume that the temperature of the hot water entering the tower is 120°F.

Sulfates in circulating water are 5 × 20 = 100 ppm SO_4

or $100 \times \dfrac{136}{96} = 142$ ppm $CaSO_4$

$CaSO_4$ solubility limit = 2,200 ppm $CaSO_4$

Additional sulfates formation permissible 2,200 - 14 = 2,058 ppm $CaSO_4$

or $2,058 \times 39 \dfrac{96}{136} = 142$ ppm as SO_4

The alkalinity in the circulation water, if not converted into sulfates, is 5 × 90 = 450 ppm $CaCO_3$.

Assume that 10 percent of the alkalinity is left unconverted to avoid corrosion, then 450 × 0.9 = 450 ppm $CaCO_3$ → $CaSO_4$

Sulfate formed $405 \times \dfrac{96}{100} = 397$

where

136 = molecular weight of $CaSO_4$
96 = molecular weight of SO_4
100 = molecular weight of $H2SO_4$

Thus, the sulfuric acid concentration required is $405 \times \frac{96}{98} = 397$ ppm.

The composition of blowdown water in this case will be:

Hardness, ppm	$CaCO_3$,	5×118	$= 590$
Alkalinity, ppm	$CaCO_3$,	$0.1 \times 5 \times 90$	$= 45$
Sulfates, ppm	SO_4,	$5 \times 20 + 388$	$= 488$
Chlorides, ppm	Cl,	5×19	$= 95$
Silica, ppm	SiO_2,	5×2	$= 10$

In this example the cycles could have been carried much higher because we have 48 ppm of SO_4 versus 1,450 allowed.

For sulfuric acid injection, a storage drum and a proportioning pump must be provided. Carbon steel is a suitable material for the concentrated sulfuric acid drum, providing that moisture does not enter the drum. For safety purposes, it is suggested to avoid glass level gauges. It is best to install a floating-type level gauge.

The injection point of the sulfuric acid is in the pump bay, or as near as possible the to water intake. The sulfuric acid pump is normally a motor-driven proportioning pump, and an electric motor is connected to a pH analyzer installed on the cooling-water supply header so that the pump starts and stops, depending on the pH in the circulating water. Table 8-3 summarizes various chemical treating agents for cooling-water towers. An example for estimating the required amount of different chemicals follows:

Calculate the chlorine and phosphates requirements for a tower operation:

1. Chlorine. As stated in Table 3, for algae and bacterial control the normal quantity of chloride needed is 1 ppm every 4 hours daily, which represents

$$1 \text{ ppm} \times \frac{4\text{hr/day}}{24\text{hr/day}} = 0.2 \text{ppm}$$

continuously.

Suppose a tower operates with 100,000 gpm of circulating flow:

$$\text{Chorine(lb/day)} = \frac{100{,}000 \text{ gal/min} \times 1440 \text{ min/day}}{7.48 \text{ gal/ft}^3 \times 62.4 \text{ lb}}$$

$$\times \frac{0.2 \text{ lb Cl}_2}{1{,}000{,}000 \text{ lb H}_2\text{O}} = 240 \text{ lb/day}$$

2. Phosphate. From Table 8-3, the requirements of phosphate are 2-10 ppm of PO_4. The loss of phosphates will be due only to slowdown and windage. To calculate the phosphate requirements:

$$\text{Phosphate (lb/day)} = \frac{10 \text{ lb } PO_4 \times (W+B) \text{ gal/min} \times 1440 \text{ min/day}}{1{,}000{,}000 \text{ lb } H_2O \times 7.48 \text{ gal/ft}^3 \times \text{ft}^3 \; 62.4 \text{ lb}}$$

Table 8-3 Chemical Treating Agents for Cooling Water Towers

Chemical and Common Name	Water Treatment Use	Quantity (ppm in circulating water)
Inorganic Chromate Salts	Corrosion control	300 - 500 ppm of CrO_4
Inorganic and Organic Phosphates and Polyphosphates	Scale and corrosion control	2 - 10 ppm of PO_4
Chromate and Phosphate Combination Treatment	Corrosion control	10 CrO_4 + 40 PO_4 20 - 50 ppm
Lignin and Tannin Organic	Scale and corrosion control	20 - 50 ppm
Organic Chromates		5 - 20 ppm
Chlorine and Chlorinated Phenols	Algae and bacterial slime	1 ppm 4 hr/day
Quaternary Ammonium Copper Complexes	Algae and bacterial control	200 ppm intermittent
Sulfuric Acid	Solubility control	As necessary to maintain same residual alkalinity

where:

    W = windage losses
    B = blowdown losses

All other chemicals can be calculated in the same way.

PLASTIC COOLING TOWERS

Corrosion problems and costly water treatment can be minimized in many applications through the use of plastics. Since about 1970 the use of industrial-grade plastics has become widely accepted in prepackaged, factory assembled cooling-tower units. There are numerous advantages to component construction, including polyethylene shell, ABS wet decking and drift eliminator system, and PVC distribution assembly, which have proven superior to steel and wood construction in many applications. There are several advantages of plastics construction over wood or steel. Plastics are noncorrosive, have a seamless, leakproof one-piece shell, are nonbrittle, nonporous/one-piece when wet and are lighter. Further, they require less maintenance and give longer service.

Each tower consists of a seamless tubular shell, which houses a specially designed wet decking. The decking, of angled baffle construction and wound in a continuous spiral, provides good air/water contact. These are counterflow operating towers. The wet decking and shells are constructed of noncorrosive plastics, which are impervious to industrial smoke, chemical fumes salt, heavy dust, and alkaline, chlorinated, or acid water. In addition, these materials resist algae growth, which greatly reduces water treatment chemical costs and maintenance.

Although advanced designs for ice prevention systems have been developed, ice-related structural damage continues to plague cooling-tower systems, particularly hyperbolic natural draft towers. Ice damage can be a severe problem, with the potential for structural damage. Because of the grave operating problems and their frequency of occurrence, a separate section is devoted to this subject.

There has been a trend in moving away from constructing natural draft towers in geographical regions where winter climates are severe. Counterflow natural draft towers are generally less susceptible to ice damage. In these systems, the fill is sheltered within the tower shell and the air inlet design usually does not include louvers. Advanced designs and new guidelines have resulted in a significant reduction in the frequency and extent of ice damage in counterflow natural draft cooling towers.

Overall Ice Prevention System Design

There are basically three phases or subsystems currently used for ice prevention in natural draft cooling towers:

- Fill bypass systems, which are capable of diverting the entire hot-water flow directly into the tower basin, comprise the first.
- The second comprises designs that include an ice prevention ring that distributes a portion of the total hot-water flow across the cooling-tower air inlet as a veil of falling water.
- A fill zoning subsystem designed to divert the water flow away from the center of the tower fill is the third system. This creates an annular flow operating configuration with a zone of high-density water loading about the outer region of tower fill.

Each of these subsystems can be controlled by adjusting the water flow. This action can limit ice formation over a wide range of heat loads and ambient environmental conditions. Each subsystem is independently controlled and altogether they represent the best technology presently available for ice prevention. Cooper and Vodicka developed an empirical model for describing these subsystems, specifically their thermal characteristics. We will use some of the qualitative predictions of their model to describe each process in detail. Before beginning, there are three new terms we must introduce to our cooling tower vocabulary, namely, *ring water temperature*, *fill water temperature*, and *basin water temperature*.

Ring water temperature is defined as the average temperature of the water discharged from the ice prevention ring after cooling by the inlet air. Fill water temperature is the average temperature of the cooled water exiting the tower fill system. Finally, the basin water temperature is the average temperature of the water discharging from the tower's collection basin. In normal tower operation there is no water flow through the ice prevention ring and fill bypass, in which case the basin water temperature and fill water temperature are the same.

For counterflow natural draft towers under normal operating heat loads and water loadings, minimal ice formation can be expected in moderately cold environments. Figure 8-1 illustrates the variation in basin water temperature with inlet air wet-bulb temperature at different heat loads and water loadings. The plot illustrates that at conditions of full water flow and at least load, the potential for freezing does not occur until very low wet-bulb temperatures are reached. (Note that since water temperature gradients exist between the central and peripheral regions of large cooling towers, a basin or fill water temperature of 40°F or less constitutes a condition of high freezing potential in a cooling tower.) A more subtle observation that can be made from Figure 8-1 is that a reduction in water loading significantly increases the tower's susceptibility to localized freezing. By reducing the water loading to the fill, the tower impedance of air flow is reduced. This results in an increase in the air mass flow rate through the tower. High water loading over the fill is most often recommended for winter operation.

Cooling Waters   261

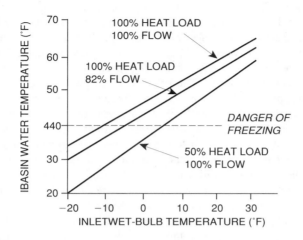

Figure 8-1 Illustrates the danger of freezing for normal cooling-tower operation

Mechanics of the Fill Bypass and Ice Prevention Ring Sections

In normal operating mode, the entire water loading is distributed over the heat-exchange surface (fill material). In the fill bypass operating mode, a portion of the hot-water loading is diverted directly to the cooling-tower basin. This results in an increase in the overall average water basin temperature. Fill bypass operation is an essential step when placing cooling towers onstream during cold weather.

In the northeastern region of the United States, a great many natural draft cooling towers are operated in a fill bypass mode and, in many cases, this represents the only means of controlling tower water temperatures in cold weather. At a given wet-bulb temperature, as more water was made to bypass the fill, the basin water temperature was observed to increase, while the fill water temperature decreased. The decrease in fill water temperature can be attributed to the decrease in the fill water loading and, thus, the subsequent reduction in total thermal energy available to the fill. Under steady-state conditions, the bypass operating arrangement can maintain elevated basin water temperatures in moderately cold weather.

There is, however, a danger for ice formation in the fill if too much water is bypassed. In the majority of cooling-tower operations, standard practice is to open and close the bypass valves in a cyclic fashion to maintain a desired average basin water temperature while minimizing fill ice formation. The main disadvantage with bypass cycling is that continual operator attention is generally required.

The portion of the tower most susceptible to icing is at the air inlet, which includes the diagonal structural supports and peripheral fill sections. This is especially true with counterflow natural draft towers. In recent years, the so-called *ice prevention ring* has been incorporated into cooling-tower designs operating in colder climates. The basis for its design is that for any given wet-bulb temperature, as the ring water flow is increased, the fill water, ring water, and basin water temperatures increase. Modern ice prevention rings operate typically with ring flows of 20 percent to 40 percent of the total water loading to the tower.

In general, the thermal effects of the ice prevention ring water on the average fill water temperature is small. The role of the ice prevention ring is to preheat the air entering the fill and thus prevent the formation of ice in the peripheral fill sections of the tower. The falling veil of water is densest at the upper regions of the air inlet. This characteristic causes a constriction at the air inlet and causes the air stream to change its direction toward the lower portion of the air inlet. The peripheral sections of the fill thus become effectively shielded from direct exposure to the incoming cold air. When the ice prevention ring and bypass system are operated simultaneously, a wide range of ice-free operations can be achieved by proportioning the proper water flows to each subsystem. Also, by increasing the bypass flow rate, increases in the ring water and basin water temperatures are achieved, accompanied by a decrease in the fill water temperature.

Another operating mode for the fill bypass and ice prevention ring is illustrated in Figure 8-2. In this case, the hot-water loading over the fill is totally diverted to the ice prevention ring and fill bypass subsystem. In normal operation this is accomplished by opening the bypass valves to the point that the water level in the distribution flumes falls below the inverts of the fill water distribution piping. At this lower flume water depth, no water reaches the fill section, causing a substantial head to exist over the open valves connecting the flumes to the ice prevention ring subsystem. This is a preferred operating mode in extremely cold weather as it provides steady-state, low heat loading operation. This mode essentially short circuits the fill section, causing all heat rejection to take place in the descending water in the ice prevention ring.

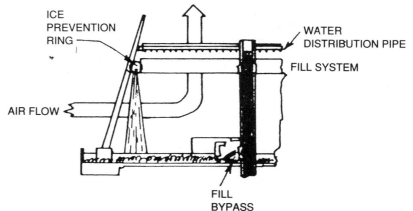

Figure 8-2 Diagram showing the proper flow allocation during low heat load operation, with water flow diverted from the fill section

There is one final ice prevention configuration worth noting, only because it is still utilized in older installations. In this design, a circumferential hot-water distribution pipe is positioned above and adjacent to the inside face of the tower air inlet. This subsystem is referred to as a deicing ring and consists of multiple-pipe sections that are perforated or stalled on the underside. The design includes valve-controlled flow connections to the main fill hot-water distribution system. Typical diameters for the deicing ring pipe section are between 6 and 12 inches, with maximum design flow about 10 percent of the total water loading.

The deicing ring was designed to distribute a small amount of hot water over the air inlet opening, thus causing ice accumulations on the air inlet structures to melt. In general, industry has complained that the deicing ring does not effectively minimize peripheral ice accumulations. In fact, numerous installations have been reported to freeze and rupture, and motor-operated deicing flow control valves have been proven to be unreliable because of freezing and corrosion. This approach is no longer applied to new natural draft towers but can be found on installations that have been in service more than ten years.

Supplemental Ice Control: Fill Zoning

The fill zoning subsystem is another supplemental approach for ice formation control. In this operating mode, the hot water loading over the center region of the tower fill is diverted to the peripheral fill sections. This increase in the peripheral fill region water loading causes a dramatic reduction of the effective interfacial contact area between the air and water. Figure 8-3 illustrates a typical fill water distribution pattern in the zoned mode of operation. The annular air flow formed results in an increase in the air flow impedance of the tower, thus reducing its air loading through the fill section. Fill zoning alone unfortunately is not enough to protect the peripheral fill sections from ice damage. The best available technology recommends operating counterflow systems with the zoning subsystem in conjunction with the ice prevention ring.

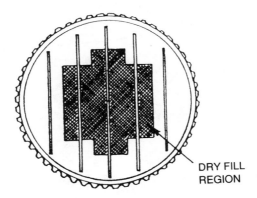

Figure 8-3 Typical fill water distribution pattern in the zoned mode of operation

GUIDELINES FOR INTEGRATED SYSTEM OPERATION

There are many applications in which it is important to maintain an average basin water temperature at an optimum value. One example is a utility cooling tower application in which an optimum average basin water temperature is required to ensure efficient turbine operation. In this example, the optimum temperature falls between 60° and 75°F. Ice prevention systems should be designed to provide sufficient flexibility to control the basin water temperature between specified limits without significant ice formation for a wide range of heat load and ambient environmental conditions.

This flexibility can be achieved through the use of three operating nobs, that is, by proper adjustment of water flow allocations to the three ice prevention subsystems. We can summarize everything into four operating modes.

- Mode I Operation is the normal cooling-tower operating fashion in which hot water is distributed evenly over the entire fill plan area. In this operation, all the valves to the ice prevention ring are in the closed position. The fill bypass can be operated if needed to maintain an optimum basin water temperature.
- Mode II Operation comprises full operation of the ice prevention ring to prevent icing on the peripheral fill sections. The fill bypass subsystem can be operated to provide a specified temperature range for the basin water.
- Mode III Operation is the one in which the tower is zoned and the ice prevention ring is fully activated. Again, the fill bypass can be operated within specified limits.
- Mode IV Operation functions with the hot-water flow to the tower entirely diverted to the fill bypass and ice prevention ring. Figure 8-4 summarizes the thermal performances of the four ice prevention operation modes.

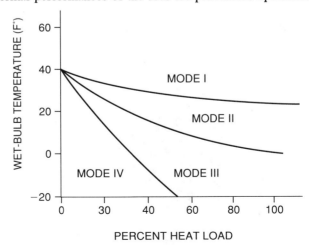

Figure 8-4 Operating regions for the winter operating modes for ice prevention

Index

acid wastes 37, 81
aeration 122
aerobic 39
algae 116
alkalinity 95
amines 38
anaerobic 39
aquifers 172
artisan aquifer 16
bacteria 44, 85, 86, 109, 233
bacteria coliform group 109
bacterial examination 44, 85, 86
biological monitoring 202
blowoff 219
boiler feed water 206, 209
bubble aerators 123
calcium carbonate 231
cathodic protection 223
chemical industries 34
chemical properties 64
chemical treating 258
chlorination 162, 208, 250
clarifiers 128
coagulation 43, 96, 124
cold lime soda 214
colloidal matter 59, 71, 77
color 86, 208
concentration 75
condensate 217
confined aquifer 174
contact time 131
contamination 174
cooling water 207, 220, 236, 243, 251
copper sulfate treatment 250
corrosion 219, 222, 242, 245

dechlorination 161
deep wells 18
demineralization 164
density 59
deration 216
diatoms 116, 153
digging of wells 2
disinfection 155
dispersions 70
dissolved gases 57
dissolved solids 57
distillation 215
distribution 24
drilled wells 18
drinking water 46, 184
dual-media filters 148
dyeing 32
electrolytes 81
emergency storage 25
emulsions 80
equalization 24
eutrophication 39
evaporation 7, 54
evapotranspiration 9
fermentation 109
filtration 124, 143
fire storage 25
flocculation 130, 132
flotation 130
flow rates 205
fluoridation 162
fluoride 97
foaming 35
food products 34
freezing mixtures 55, 61
fresh-water supply 6
fungi 116
gas solubility 63
granular media 139
groundwater 15, 20, 29, 167
groundwater biology 21
groundwater monitoring 180
groundwater profile 194

hardness 41, 42, 99
heat capacity 53
hotels 221
human factors 39
hydrate formation 66
hydrogen cation-exchange process 213
hydrogen sulfide 38, 209
hydrogen-ion concentration 107
hydrogeologic setting 193
hydrologic cycle 5, 6
hydrologic pathways 187
hydrolysis 67
ice control 263
impurities 3
industrial water requirements 204
infiltration 7, 177
inorganics 51
ion exchange 164, 213
ionization 81
iron 34, 209, 233
iron and manganese 101
irrigation 8, 30
latent heat 63
Lavoisier 1
leachate 190
Legionnaires' disease 253
manganese 101, 209
marble test 103
measurement units 83
membrane filter 112
metal oxides 64
methane 38
microorganisms classifications 116
microscopic examinations 3, 116
mineral analysis 59
monitoring 191, 199
mud balls 104
multimedia filter 149

municipal supply 30
natural lakes 13
nonelectrolytes 81 normal solutions 77, 78
odor 86, 208
office buildings 220
organic growths 252
organic pollutants 38
osmotic pressure 72
paper making 32
pathogenic organisms 161
pH scale 108
phosphate treatment 218
physical properties 50, 59
plastic cooling towers 259
plume of leachate 196
pollutant categories 37
potable water 43
precipitation 12
pressure filter 153
pressures 205
pretreatment procedures 246
process water 207
properties 40
protozoa 116
public health 41
purification 120
quality 26, 40
quality standards 28
radial flow filter 151
rainwater 12
rapid filters 139
rapid-mix units 130
rapid-sand filter 141
rate of corrosion 228
recharge 170
rechlorination 161
relative concentrations 68
removal of oxygen 232
representative samples 85
reservoirs 29

residual chlorine 105
rivers 13
runoff 7, 12
safe drinking water act 178
safe yield 13
saline water 37
salt pollution 37
salts 81
sampling 85, 200
sampling frequency 85
sampling procedure 86
sandstones 20
sanitary analysis 58
sanitation 14
saturated solutions 72
scale 239
scale in boilers 36
sediment 242
sedimentary rocks 20
sedimentation 124
sedimentation tanks 129
settling 123
settling phenomenon 126
shallow driven well 17
shallow dug well 16
silica 215
site inspection 192
slow-sand filters 142, 154
sludge blanket clarifier 136
soap manufacturing 34
sodium cation exchange 212, 215, 248
soft waters 49
sole source aquifers 179
solubilities 74
solutions 69, 71
sources 3, 1, 27
sources of contamination 176
specific heat 53, 63
spring basin construction 19
spring water 48
springs 18

standard plate count 114
steam generation 35
storage 25, 205
stream flow infiltration 7
structural monitoring 201
subsurface water 15
sulfite treatment 218
surface runoff 7
surface water 12, 29, 47
surface water supplies 10, 13
suspended impurities 123
suspended solids 59
suspensions 71
tanning 33
taste 90, 208
temperature 60, 84
textile industry 30
transpiration 9
treatment of cooling-water 244
true solutions 71
turbidity 208
turbidity test 93
unconfined aquifer 173
underground water zones 169
universal solvent 60, 72
unsaturated flow 8
upflow filter 150
uses of water 22
vapor pressure 55, 56, 57
wastewaters 145
water composition 46
water conditioning 206, 220
water consumption 21
water demand 22
water distribution 23
water impurities 3
water quality 11, 121
water softening 42
water supply 3, 23, 58
water treatment 40
water withdrawals 169
zeolite process 212